SOUTHERN · TIMES ·

Contents

Introduction	
Original colour around Nationalisation	5
The South Eastern from Ashford to Dover and coastal problems for William Cubitt. Part 2 — Jeremy Clarke	9
Cycling on the Southern — Ian Shawyer	15
The SECR 'J' class tank engines	19
By Withered Arm from Barnstaple Junction to Halwill Junction via Torrington	27
Farnborough Air Show Traffic. From the notes of John Davenport	45
Stephen Townroe's colour archive: The 1952 Shawford derailment	52
In and around Fratton — Images by Tony Harris	60
The Elham Valley line — A Earle Edwards	63
Treasures from the Bluebell Railway Museum — Tony Hillman	67
The 1948 Southern Region Locomotive Building Programme — The nightmare ...	70
From the Footplate	78

The next issue of Southern Times, No 10, will contain:
The Mid Sussex Line
Demolition at Crystal Palace by R C Riley
The Southern main line diesels
Moving stock to Droxford in 1967
From the Footplate etc etc.

The Transport Treasury
TIMES SERIES

SOUTHERN TIMES

Front Cover: Variation on the standard BR black livery. 'L' class 4-4-0 No 31771 probably photographed at Canterbury West soon after overhaul/repaint and not long after nationalisation. This is one of a small series of Dufay colour images recently squired by the Transport Treasury; one, that of an N15X at Waterloo was reproduced in Issue 8, the others are seen on the cover of and within this issue. Dufay colour was marketed between 1933 and 1958 and although using a different process to later colour film could also give reasonable results. Here the main point of interest has to be the bogie wheel centres still in SR green. Might it have been the simple case of a bogie swap in works from an engine still in the older livery? (Further Dufay colour images appear above and on pages 4-7 of this issue.) *Transport Treasury*

Above: Another 'L', this time No 31775 awaiting the off. The location is Ramsgate (thanks to Mike King for this and other locations!).Regretfully though no date. The 'weathered' Southern livery is in the opinion of the writer not unattractive. *Transport Treasury*

Rear Cover: We did intimate last time there would be various R C Riley colour images on the way. Here is one of the then three surviving members of the C14 class recorded at its usual place of work, the sidings near the Royal Pier, Southampton on 26 June 1957. An article on these little engines and the associated S14 type appeared in Southern Times issue 8. *RCR 3153 / Transport Treasury*

Copies of many of the images within **SOUTHERN TIMES** are available for purchase/download.

In addition the Transport Treasury Archive contains tens of thousands of other UK, Irish and some European railway photographs.

© Kevin Robertson. Images (unless credited otherwise) and design The Transport Treasury 2024

ISBN 978-1-913251-77-2

First Published in 2024 by Transport Treasury Publishing Ltd.,
16 Highworth Close, High Wycombe HP13 7PJ

www.ttpublishing.co.uk or for editorial issues and contributions email to southerntimes@email.com

Printed in Malta by Gutenberg Press.

The copyright holders hereby give notice that all rights to this work are reserved.
Aside from brief passages for the purpose of review, no part of this work may be reproduced, copied by electronic or other means, or otherwise stored in any information storage and retrieval system without written permission from the Publisher.

This includes the illustrations herein which shall remain the copyright of the respective copyright holder.

INTRODUCTION

Welcome to Issue 9 of Southern Times, I genuinely hope you enjoy what follows in the next 78 pages.

Compiling 'ST' is a pleasure and I promise it is not a question of what do I put in but more what do I leave out! That said it is also all too easy to include, unintentionally, personal favourites and that might even be said with the article on pages 70-77 but I would justify this - and that does not mean make excuses - for it is new information and as such can only add to our field of knowledge.

I expect too that like me, you read other magazines and journals and not all railway related. If it is a publication dealing with the current scene, be that railways, computers, even the television schedule, then the compilers are able to produce material that is new and unknown. A historic publication is different especially one where we have to attempt to appeal to a broad church.

Yes, we could include copious reports of staff movements, engine names and numbers, train services and timetables and the like; basically material which is of a fringe interest, or little interest at all simply because it is already well known. In short how many articles are possible on say the 'Merchant Navy' class that will tell us something new?

Sorry if that sounds cynical, it is not at all meant to be and I am certainly not trying to talk myself out of a job, instead what I am saying is there remains a fine balance between repeating history hopefully in a new form and simple plagiarism. An example in this issue is the piece on the Elham Valley line in Kent. I think it unlikely anyone will ever usurp Brain Hart's superlative writings on this short line but with the addition of the information contained in the 'Southern Railway' magazine for 1947 we can hopefully add a little more to the history. Be assured, we will always acknowledge others' work.

Which leads on ('in' perhaps might be a better term) to the SR Magazine and other contemporary magazines and books. I suspect that like me you have 'a few' railway books on the shelves of your own library. Personally at the last count the number was around 2,500 - we keep them downstairs....! Some are modern, some relate to specific topics and are perhaps the standard reference works, and others simply because I like them. A number have also been pushed to the back in recent years, 2,500 takes up a lot of room and I am sure I am not alone when I say certain volumes may get covered by others, only to be rediscovered months or even years later.

This was the type of discovery I made recently with a few volumes of short lived magazines and books published many, many years ago. The surprise was they can produce much useful material and we will be digging into their pages in future issues.

We have also often said we welcome suggestions for articles, things you would like covered. One of these came from a recent correspondent who commented he would like to see something on the Mid Sussex line. Well not in this issue but look out for No 10 - we aim to please.

'Southern Times', like its sister 'Times' publications produced by Transport Treasury, is about unashamed nostalgia. With the pictures and articles we are genuinely trying to 'put you there', to recall the sound of the announcer, the anticipation of the train, the spring in the seats, the colours, the moquette, the smells, in fact everything that made up the railway we all knew.

As I write this the July General Election is only about a week away. It may mean changes to the current privatised railway network, although sadly I fear without a return to semaphore signals, 4COR sets and Bulleid rolling stock.

I will admit I never did heed of the cautionary tale from a friend 40 years ago who commented 'you should be noting and recording today's scene as it will change all too quickly'. He was right - and I never did. Instead I still long for those halcyon days referred to in the previous paragraph, unashamed nostalgia perhaps and something we try and produce in each issue of 'Southern Times'.

Kevin Robertson - Berkshire 2024.

Top: No 796 *Sir Dodinas le Savage* (we think - the number is not 100% clear on the cabside) photographed leaving Broadstairs. As with the other images in this selection we regret we have no dates and no original caption information.

Bottom: A very clean 'C' class No 31112 pictured at Canterbury West. When clean even BR plain black could look attractive.

Opposite top: No 1770 engaged in a spot of shunting, station pilot perhaps, at Ashford.

Opposite bottom: 1903 built 'D' class 4-4-0 SR No 1748 at an unknown location. This engine was withdrawn fairly soon into BR ownership in March 1951.

Original colour around Nationalisation

This time more definite identification of this 4-6-0, No 763 *Sir Bors de Ganis* at Dover. Between 1948 and 1948 this engine was at Stewarts Lane, Ramsgate and then back to Stewarts Lane before moving to the Western section in 1955 (Basingstoke, Eastleigh, Bournemouth and Nine Elms) never to return to its old haunts. After a service life of 35 years, 4 months and 1 day, it was withdrawn in September 1960.

No 1770 again, this time at Ramsgate. This machine remained on the Eastern Section at Dover and then Tonbridge until 1952 before migrating west to Nine Elms. Interesting that it is coupled complete with vacuum brake hose to what is almost certainly a 'Schools' behind, although look closely and a member of staff is seen between the two engines. With grateful thanks to Mike King for assisting in identifying locations.

An unidentified 'HA' (later Class 71) Bo-Bo electric passing Folkestone Warren with the down 'Golden Arrow' in October 1963.
Graham Smith courtesy Richard Sissons.

The South Eastern from Ashford to Dover and coastal problems for William Cubitt
Part 2
Jeremy Clarke

After some consultation with the Board of Trade it was deemed impractical to recover the train until after the war, reopening of the line thus being delayed until 11 August 1919. During that four-year closure drainage tunnels were bored into the face of the chalk in an attempt to halt erosion, though this was not wholly successful. Indeed, another slip occurred in 1939 but on nothing like the scale of the one a quarter of a century earlier.

More far-reaching works were undertaken from 1948 in the form of massive 'toe-weighting' along the shoreline, more substantial drainage channels being bored into the base of the cliff, and protection from erosion by the sea being given to the clay outcrop. Extension of sea defences eastward almost as far as Abbotscliffe* tunnel were undertaken between 1953 and 1955. A halt for staff who maintain the area is in place at milepost 72 from Charing Cross. This was originally built in 1888 to bring visitors to this unique landscape but as Board of Trade authority had not been sought for it, opening was delayed until 1908. It closed for public use very soon after the outbreak of war in 1939.

A mile or so beyond what was referred to as Warren Halt, the line passed an engineers' siding controlled by the tiny Abbotscliffe signal box. Notwithstanding its diminutive size this was a block post and in use until the end of April 1961. The original box had been swept away in the 1915 landslip, the signalman on duty having a lucky escape.

At milepost 72¼ the line enters the 1 mile 182 yards of Abbotscliffe tunnel, on a downgrade of 1 in 264. Cubitt found the chalk here to be perfectly stable and tunnelled without undue difficulty. As well as the usual method of working outwards from shafts sunk on the tunnel line, horizontal galleries or drifts were driven into the cliff face from a roadway dug into the cliff side. This permitted spoil to be wheeled out and tipped directly into the sea. Despite its stability the tunnel is lined with six rings of brick.

Round Down Cliff, rising sheer some 300' above the sea formed the next obstacle Cubitt faced. Here the chalk proved to be very unstable, so much so Cubitt considered it too dangerous for tunnelling. He concluded that the safest means of getting the line through was to blast the obstruction away. Following consultation with General Pasley, the Board of Trade's Chief Inspector of Railways who approved the scheme, Cubitt handed over the problem to an explosives expert, Lieut. Hutchinson, RE. Hutchinson had charges totalling 18,000lbs of gunpowder (just over 8 tons), placed within the base of the cliff beforehand, then on 26 January 1843, the due day of firing, huge crowds gathered to watch what they believed would be a spectacular display of pyrotechnics. They were to be disappointed for, as the newspapers reported, there was no massive explosion, simply a low rumble before the bottom of the cliff bellied outwards and the great mass of chalk simply sank into the sea. To Hutchinson's credit this was probably the first occasion on which detonation had been made by electricity.

Apart from holding the Chairmanship of the SER, Sir Edward Watkin's other interests included appointment to a senior office with the Channel Tunnel Company. In 1881 it received approval from Parliament to begin construction, a pilot tunnel of eventually 1¼ miles under the sea being completed as well as another 1½ miles from the French coast before the War Office ordered its suspension**. A side advantage was that coal was discovered during the boring of the vertical entrance shaft which led to the development of the Kent coalfield. Access to the Dover colliery sidings set immediately west of Shakespeare tunnel and on the up side, was controlled by Shakespeare signalbox set almost opposite the outlet. A number of other collieries were opened in East Kent, though only a minority proved economically viable for long. Dover colliery was not among these, geological problems contributing to its closure in December 1915.

As in The Warren, a staff halt was provided from quite early days close to the western portal of Shakespeare

Part 1 of this article appeared in Southern Times No 8

With a light load behind the tender, Q1 No 33031 is seen arriving at Ashford on a down freight, 27 August 1960. *Dennis C Ovenden*

tunnel though it never had any public use. Unusually perhaps a trip wire extends this far on the down side of the line from the western entrance of Abbotscliffe tunnel. It will cause signals to be set at 'danger' should a slip disturb it.

Unlike at Round Down, the chalk in Shakespeare Cliff proved relatively stable but Cubitt was not confident enough in its stability to repeat the double-line tunnel as at Abbotscliffe. Nevertheless, he adopted the same method of disposing of much of the chalk by driving side galleries and tipping the mined material directly into the sea. Consequently here two single line tunnels were driven with a solid chalk pillar some ten feet thick between the bores. Cubitt made them of Gothic profile which lessened the pressure on the crown of the arch, the forces now principally being inward rather than downward. The height of the tunnels is not consistent, being 26½' high at one end and 29' at the other. The bores are 1,387 yards long and on a downgrade of 1 in 240.

Originally the line left the tunnel on a timber trestle viaduct along the beach, the piles being sunk down into the underlying chalk. Not long after its formation, the Southern Railway began to fill the spaces in and around the piles with chalk and widen the whole formation for additional track behind a massive sea wall, work being completed in 1927. Nearly 90 years later at Christmas 2015, large cracks over a distance of up to 250 metres were seen in the concrete causing the line east of Folkestone to be closed immediately while a full-scale investigation was carried out. Among other things, this investigation found that records of exactly what the Southern had done here were particularly scant. More importantly it showed the level of Shakespeare Beach had been eroded by recent storms by about two metres, exposing the wall's foundations and creating sink holes beneath them. It was also suggested the toe-weighting and the extensive concrete aprons pushed out from the foreshore at The Warren in connection with it may have caused or at least influenced a more destructive change in tidal flow along this part of The Channel.

Whatever the cause, the line remained closed for nine months while, in a reflection of Cubitt's ideas perhaps, a new reinforced concrete viaduct some 235m long was erected behind a wall. These works, however, were faced by strong sea defences composed mainly of thousands of tons of rock.

No 34083 *605 Squadron* passing Sandling with a down train, 4 April 1960. The route discs would indicate the service originating possibly at London Bridge and destined for Folkestone / Dover via Chislehurst, Tonbridge and Ashford. *Dennis C Ovenden*

The short tunnel under Archcliff Fort, now a Scheduled Monument and 76 miles into the journey from Charing Cross, was opened up by the Southern Railway in 1928 when improved loco facilities were required locally, part of the adjacent foreshore being reclaimed at the same time.

The SER's Dover Town terminus lay immediately east of the tunnel with a small engine shed squeezed into a tight angle on the down side. The Chatham had arrived at its Harbour station, due north of Town in 1861 but the double track spur between Archcliff Junction on the South Eastern and Hawkesbury Street Junction in the shadow of the Chatham's Harbour station, a line also requiring demolition of much residential property, did not open until June 1881. Its main purpose was to permit SER trains to reach the 'joint' line to Deal from Buckland Junction, an unusual and surprising piece of co-operation between the two warring companies. Meanwhile, the South Eastern had first run through on to Admiralty Pier in 1861, the LCDR following in 1864.

The Town station closed in 1914, the Harbour station in 1927, all trains into Dover thereafter - other than Continental services - using the ex-Chatham's Priory station sited a half-mile north of Harbour. It was the construction of the Marine station against the eastern face of the Admiralty pier that provided the greatest evidence of co-operation between the two under the aegis of the SE&CR. Work began in 1909 with the creation of a sea wall and then the infilling of the area of about twelve acres thus created to provide a solid base for an impressive station to welcome Continental visitors. Construction of the buildings started in 1913 but the station was still incomplete when War broke out in July 1914. Ferry sailings had long been halted by the time construction was far enough advanced for the military to take over the station on 2 February 1915. It was not returned for public use until January 1919. The station again saw much military action during WW2, particularly during the Dunkirk evacuation. It also suffered damage by both shelling and bombing as did the Loco depot and Dover town in general.

D2399 shunting vehicles at the Dover train ferry terminal in August 1964. *Graham Smith courtesy Richard Sissons*

Dover Marine became Dover Western Docks in May 1979 but with the passage of the Channel Tunnel Bill and the start of boring at the end of 1987 it could not survive. The end came on 25 September 1994 though not quite, for some trains worked into/out of it for another two months, all traffic being withdrawn on closure of the signal box in July 1995.

Locomotive work over the SER was restricted for many years by Edward Watkin's insistence on speed being limited to a maximum of 60mph, though there is plenty of evidence suggesting enginemen were often not so constrained. It may also be noted that the South Eastern considered Folkestone to be its main port for the Continent so records of loco work between Folkestone and Dover in both SER and pre-war SECR days are in a minority. O S Nock has published several logged runs made by J Pearson Pattinson, a regular train timer from the late years of the 19th century. He favoured the Chatham as a railway but that did not affect the neutrality of his recordings.

At the time Pattinson was making his travels, most of the major SER main line work was done by James Stirling's excellent 'F' class 4-4-0s. To modern day engineers these would seem grossly short of boiler capacity for engines with 19"x26" cylinders, for the heating surface was only 985 sq ft, the grate area but 16½ sq ft and the pressure 160psi. For a line with some long and demanding gradients, particularly the later 1868 mainline through Sevenoaks, the 7'0" driving wheels would also seem unsuitably large. Nevertheless, the engines must have steamed freely to put up the day-to-day performances Pattinson noted. No 54 for example, with 200 tons on the drawbar, having averaged just under 54mph from Tonbridge to Ashford then took only 9½ minutes to cover the 8.1 uphill miles to Westenhanger, not far short of that average at 51.2mph. The train reached Folkestone Junction in a further 7m 40 seconds, thus averaging 52.5mph down the hill pass-to-stop.

No 156, bringing a 280-ton Dover train down from London, matched No 2 for time on the 'racing stretch' from Tonbridge despite the additional load. Then, having 'slipped' at Ashford, it took the same 200 tons onward to Westenhanger at a slightly quicker average of 52.35mph, went on to average 57.4mph to pass

Folkestone Junction and reach Dover Pier a few seconds over a further six minutes, a 55.7mph overall pass-to-stop average for the 20½ miles from Ashford.

An early 20th century run logged by Charles Rous-Marten saw work by one of Wainwright's majestic and nearly-new 'D' class 4-4-0s, No 734, on a 'Folkestone Special Express' leaving Cannon Street at 4.36pm. The engine, loaded to 250 tons, was through Tonbridge in just over 40½ minutes from the start – which included PWS restrictions at Chislehurst and Sevenoaks – but then achieved a rather disappointing average in view of those delays of only 54mph across The Weald before 'slipping' three vehicles at Ashford, passed in 70¼ minutes from London. (No note is made of the weight of the 'slipped' coaches.) But the engine then averages 55½mph 'up the hill' and takes only six minutes and a few seconds more to the stop at Folkestone Central, a 57mph pass-to-stop average. The net time was calculated as two minutes less than the actual 85 minutes for the 68.7 miles. No schedule is attached, though in view of the train's title it is likely to have been the 'standard' eighty minutes.

The later and bigger 'L' class 4-4-0 No 761, a class incidentally with a general reputation for finding the 80-minute Folkestone Expresses beyond them, was charged with taking the 220 tons gross of the 4.5pm from Charing Cross down to Folkestone in that time. The run across The Weald was spoiled by a PWS at Paddock Wood but the driver had got 1½ minutes in hand by Tonbridge and was thus through Ashford a quarter-minute early at 65mph. The climb to Westenhanger reduced this to 53mph and arrival at Folkestone was that quarter-minute to the good in a net of 78 minutes.

Folkestone also tends to predominate for up direction logs and most of those published are made by Cecil J Allen. Two heavy boat train trips are timed from the junction sidings, in both cases with a Wainwright 'E' class 4-4-0 at the head. Unfortunately details are rather limited but there is sufficient information to get a flavour of the work done. No 176 has the 9.5pm to Charing Cross loaded to 325 tons gross. The all-but seven miles from the sidings to Westenhanger occupy 12½ minutes, an average start-to-pass of 33mph, with an actual 41mph at the station, whilst the train is through Ashford at 66½mph in a little less than 21 minutes from

Another 'L', this time No 31770 and previously seen in Southern colours on pages 5 and 7, this time approaching Tonbridge on 3 May 1958. *A E Bennett 3020 / Transport Treasury*

the start. By contrast No 179 is saddled with a gross load of 350 tons and takes a full minute longer than No 176 to pass Westenhanger, though the average is still a respectable 31mph.

Another 'E' class, No 315 has, coincidentally, 315 tons on the drawbar of a post-war boat train to Victoria. This run is to a schedule of 110 minutes from Dover, though the allowance onward from Folkestone Central is 96 minutes to take into account the limited speed permitted through The Warren following the reconstruction and stabilisation work there after the 1915 'slip'. A minute is cut from the fourteen-minute allowance from the Dover start to pass Folkestone Central but the following uphill 6.8 miles are taken at an average of 36.4mph suggesting Westenhanger was passed at around 47mph. The eight downhill miles to Ashford were covered in about even time, the engine being a half-minute early there.

Historically, the intermediate stations between Ashford and Folkestone Central with, perhaps, the exception of Sandling while the branch to Sandgate remained open, have always been treated rather disdainfully. Only post-Second World War when summer holidays boosted weekend traffic were they seeing any sort of frequency. The present weekday off-peak service, franchised to South Eastern, continues to reflect that.

At the time of writing just two trains run hourly each way between Ashford International and Dover Priory. The Hi-Speed service from St Pancras calls only at Folkestone Central en route in an overall time of 28 minutes for the 21¼ miles. The stopping service, which starts its journey at Charing Cross, precedes it by thirteen minutes and takes 31 minutes to Priory, calling at all four station en route. It is a reverse story in the up direction, the Hi-Speed train preceding the stopper out of Dover by nine minutes and taking twenty-six minutes to Ashford including the Folkestone Central stop. The slow train takes the even half-hour. Any augmentation seems unlikely at present.

*Abbotscliffe appears to be spelled indiscriminately with or without the final 'e': the spelling of the cliff name itself is also fluid but usually two words, 'Abbots Cliff', though sometimes with an apostrophe before the 's'.

**Attempts to restart work on the tunnel or to begin elsewhere were always thwarted by the War Office believing it could be used by an invading force, which says little about its confidence in the military's ability to defend the relatively limited size of the bore. The Night Ferry service began in 1936 after yet another Parliamentary defeat of a Channel Tunnel Bill by just seven votes.

Bibliography and References

'History of the Southern Railway', C F Dendy Marshall, rev. R W Kidner, Ian Allan Ltd., 1963.
'The South Eastern & Chatham Railway', O S Nock, Ian Allan Ltd., 1961.
'South Coast Railways, Ashford to Dover', Vic Mitchell and Keith Smith, Middleton Press, 1988.
'Southern Main Lines, Faversham to Dover', Vic Mitchell and Keith Smith, Middleton Press, 1992.
'British Railway Tunnels', Alan Blower, Ian Allan Ltd., 1964.
'Southern Electric, 1909-1979', G T Moody, 5th ed., Ian Allan Ltd., 1979.
'Railways of the Southern Region', Geoffrey Body, Patrick Stephens Ltd., 1989.
'Southern Country Stations: 2, South Eastern & Chatham Railway', John Minnis, Ian Allan Ltd., 1985.
'An Historical Survey of Southern Sheds', Chris Hawkins and George Reeve, Oxford Publishing Co., 1979.
'Atlas of the Southern Railway', Richard Garmon & Gerry Nichols, Ian Allan Ltd., 2016.
'Railway Track Diagrams, Southern & TfL', ed. Gerald Jacobs, TRACK maps 3rd ed. 2008.
'British Rail – Main Line Gradient Profiles', Ian Allan Ltd., in collaboration with Tothill Press. (undated).
Ordnance Survey Maps 1:50000 nos 179 and 189.
South Eastern Timetables to December 2019.

Limited reference has been made to the Net, the sites 'Disused Stations Site Record' and 'The Construction Index being particularly useful.'

Don't forget, copies of the many of the images in this and other issues of

SOUTHERN TIMES

are available as downloads

Opposite page: Winchester arrival, assuming the train was punctual this would have been at 10.18am, the utility van probably an additional attachment to an existing service train (or might it even have been a special working dependent upon the number of participants?). The view is of the down platform. *This and other images in this article courtesy Ian Shawyer*

Cycling on the Southern
Ian Shawyer

Cycling might almost be seen by some as a particularly modern pastime, individual and group cyclists now regularly seen in both town and country, a considerable difference from when the bicycle was regarded more of a necessary conveyance or children's pastime. Indeed it might be said the railways were in some ways partly responsible for encouraging mass cycling, for at clocking on and off times a great mass of cyclists would emerge from the various railway works up and down the country; similar behaviour experienced at numerous factories and works in the days before general car ownership.

For those however, whose weekly toil might be office based or some other sedentary occupation, the bicycle could provide exercise at weekends and in consequence local and national cycling groups would spring up accompanied by appropriate magazines.

One of these was the magazine 'Cycling' which would organise for the cyclist, tours and events. One of these was on Sunday 8 May 1938 with an excursion from Waterloo to Winchester, the return being from Brockenhurst.

Times and prices may be made out from the rather poor reproduction of the relevant handbill although perhaps more interesting are the accompanying images on what was at the start at least of a sunny day.

SOUTHERN RAILWAY

In conjunction with "Cycling"

A NEW FACILITY

CYCLING EXCURSION TO NEW FOREST

SUNDAY, MAY 8th.

RETURN RAIL FARE THIRD CLASS 7/6 INCLUDING CONVEYANCE OF CYCLE OR TANDEM.

FORWARD JOURNEY (Rail)

WATERLOO ... ... depart 9. 0 a.m.
WINCHESTER ... ... arrive 10.18 a.m.

RETURN JOURNEY (Rail)

BROCKENHURST ... ... depart 8.32 p.m.
WATERLOO ... ... arrive 10.33 p.m.

ROAD ITINERARY—

RESTAURANT FACILITIES ON THE TRAIN.

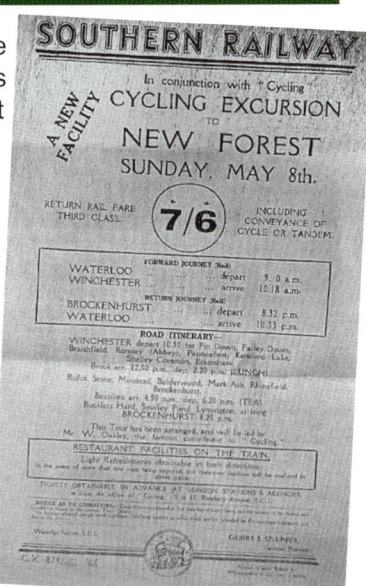

For those who recognise the route; Winchester (depart 10.30am), Pitt Down, Farley Down, Braishfield, Romsey Abbey, Pauncefoot, Kenwood Lake, Shirley Common, Bramshaw, Brook (lunch 12.50pm to 2.20pm), Rufus Stone, Minstead, Bolderwood, Mark Ash, Rhinefield, Brockenhurst, Beaulieu (tea 4.50pm to 6.20pm), Bucklers Hard, Sowley Pond, Lymington, and finally Brockenhurst (8.20pm), there are any number of steep climbs involved particularly around Farley but then those participating were probably an experienced group and not perhaps prone to getting saddle sore. Of interest too is how the itinerary and stated times read almost as if they could be part of a railway timetable.

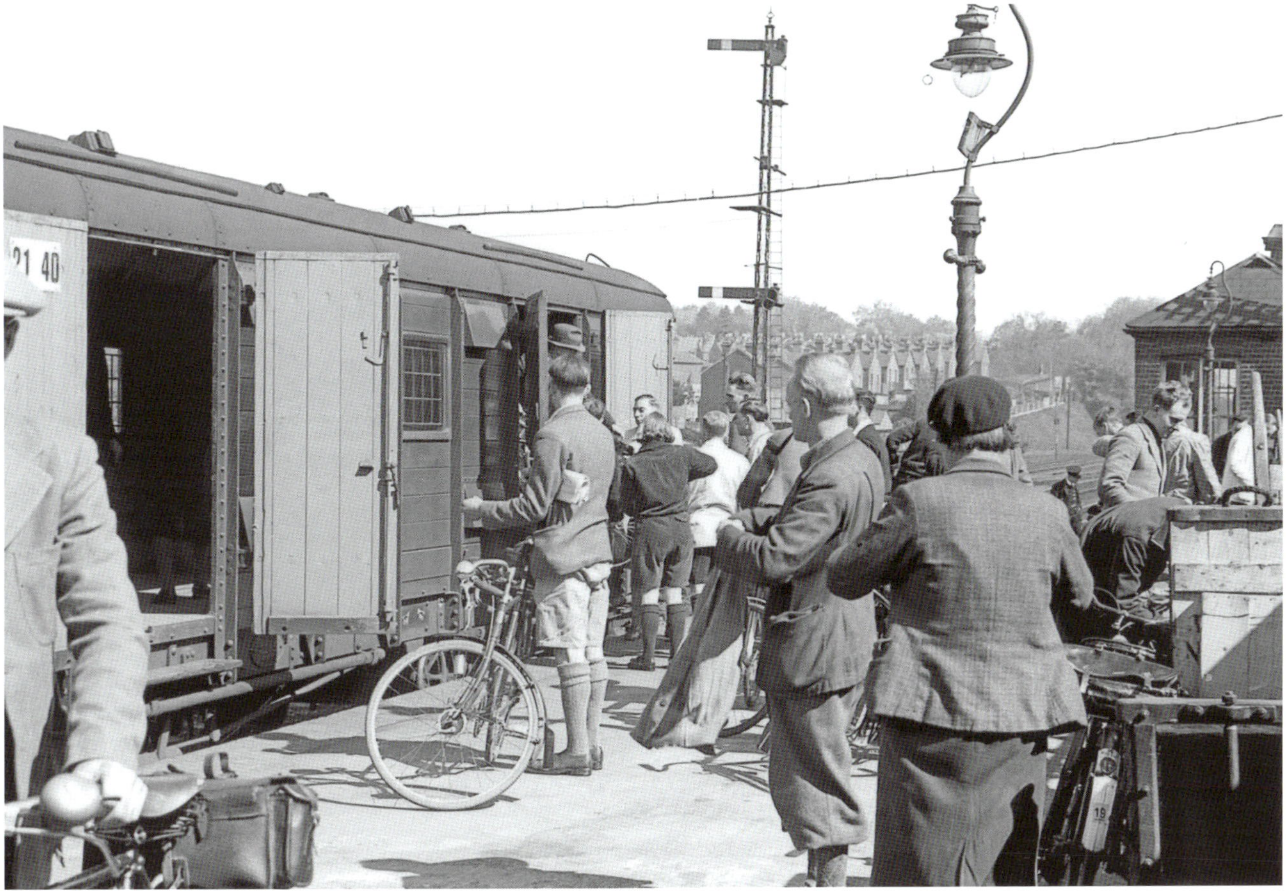

SOUTHERN TIMES

Left: The notice '1-20' does perhaps indicate there might have been more than one van in the train consist, especially when considering the number seen opposite. Fashion and hairstyles was de rigueur for the time and whilst predominantly men, there is at least one woman in the group.

This page, bottom: Supervision perhaps during the unloading? Clearly an arranged image having captured most with their heads turned.

Opposite top: Exit en masse and down Station Hill, the facade at Winchester station having changed little over the subsequent decades. Should there perhaps be a collective noun for this number, a 'pedal' of cyclists perhaps?

Opposite bottom: Away down Station Hill with the first group having turned right into City Road, the car driver perhaps overawed by the sheer numbers. Mention was made earlier of the hills to be encountered, assistance of the day for some coming from Sturmey Archer gears. Notice too the side-car and young passenger. Messrs White and Co. who were carriers for the Southern at Winchester had their offices at the foot of Station Hill whilst behind the camera on the same side of the road was the rival Pickfords organisation.

SOUTHERN RAILWAY

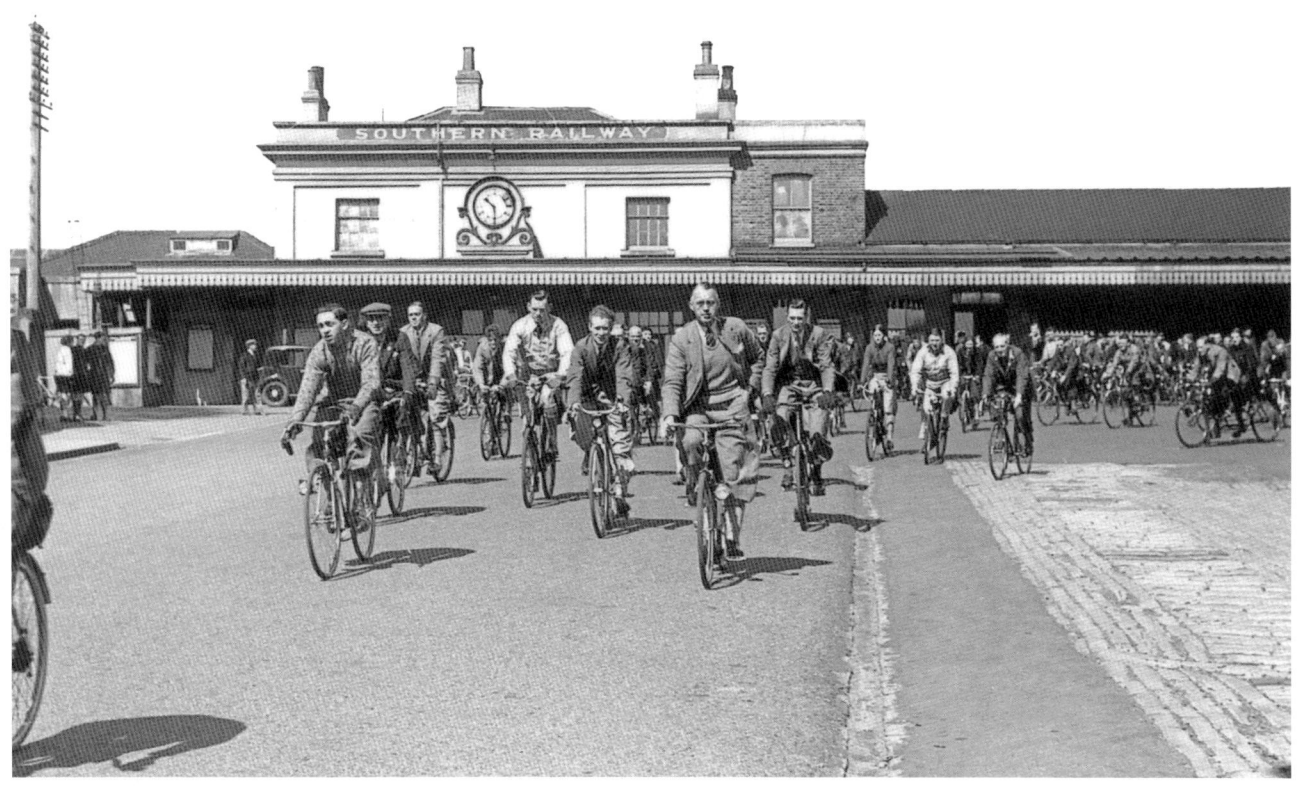

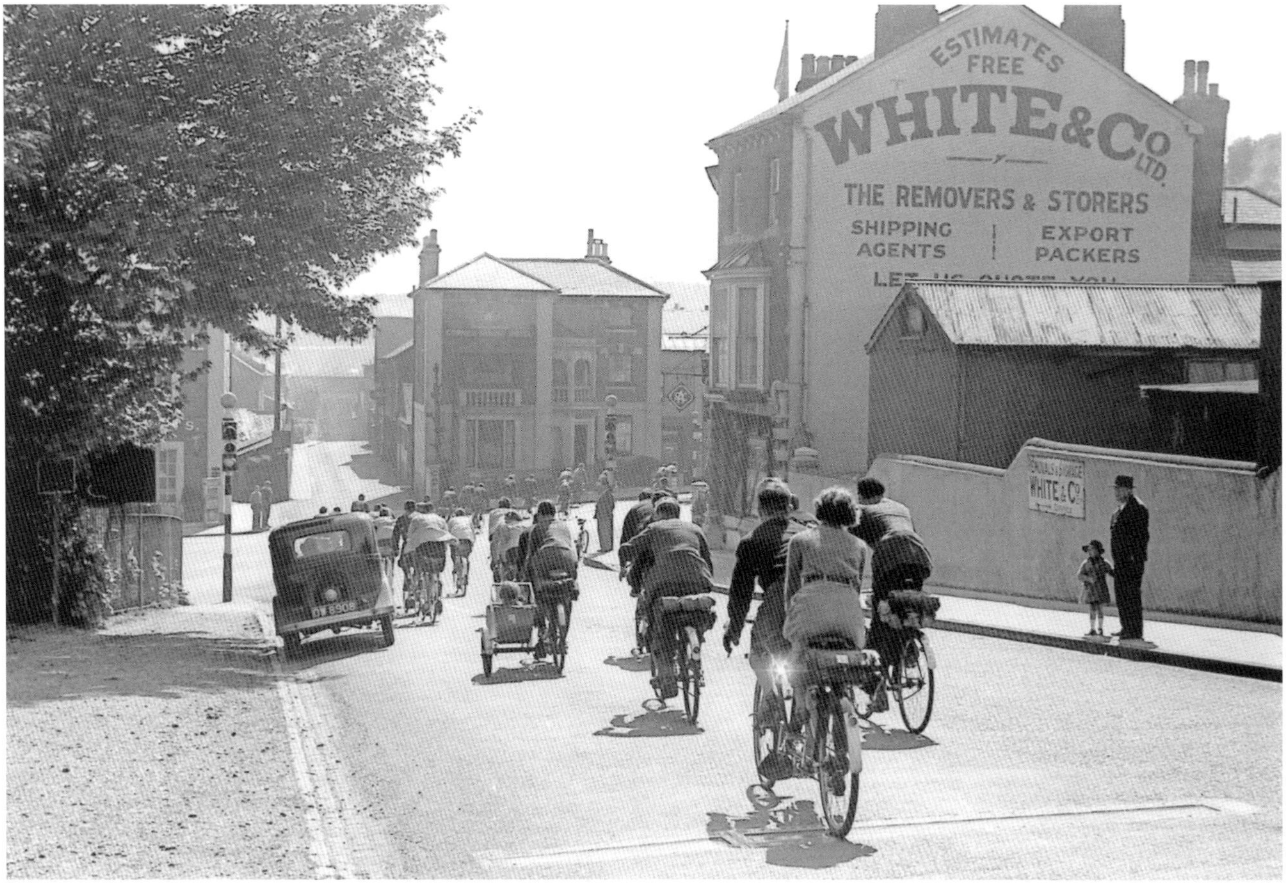
ESTIMATES FREE
WHITE & Co LTD.
THE REMOVERS & STORERS
SHIPPING AGENTS
EXPORT PACKERS
WHITE & Co

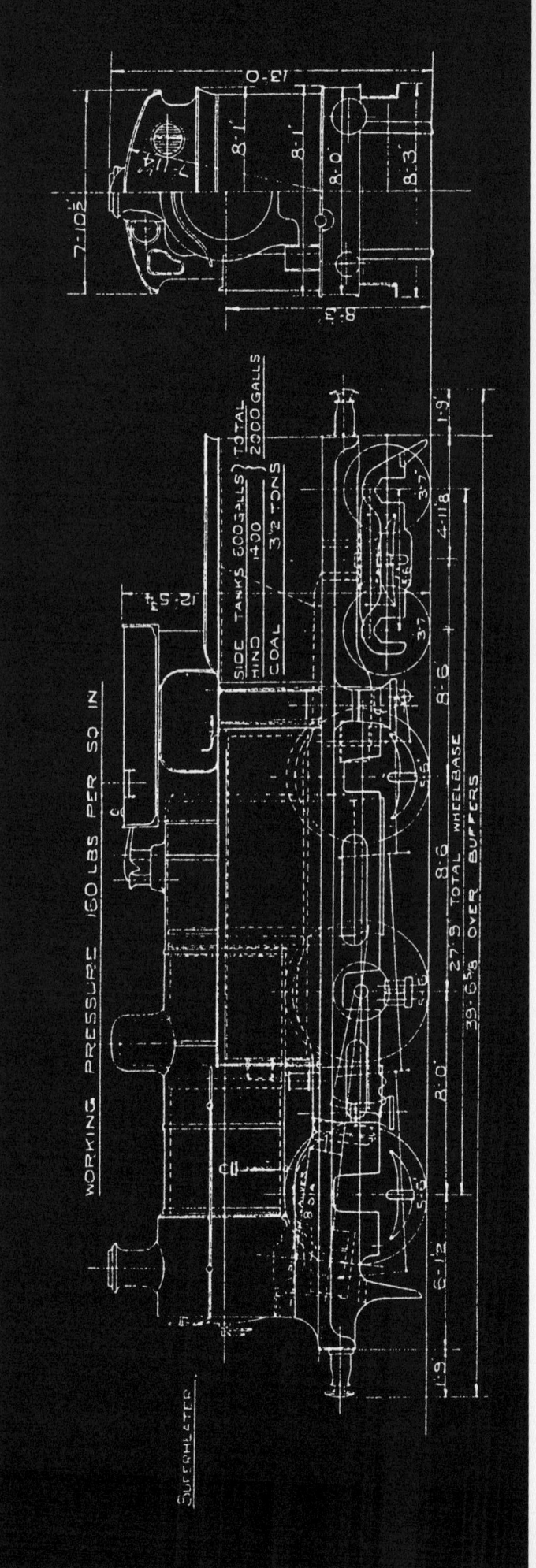

	'J' class 0-6-4T	'H' class 0-4-4T	'C' class 0-6-0
Driving Wheels	5' 6"	5' 6"	5' 2"
Bogie Wheels	3' 7"	3' 7"	n/a
Weight	70 tons 14 cwt	54 tons 8 cwt	43 tons 16 cwt
Boiler Pressure	160 psi	160 psi	160 psi
Heating Surface	1,233 sq ft	1,002½ sq ft	1,200 sq ft
Tractive Effort	20,400 lbs	17,360 lbs	19,520 lbs
Coal Capacity	3 tons 5 cwt	1.7 to 2.8 tons	n/a
Water Capacity	2,000 gallons	1,150 to 1,350 gallons	n/a
Superheater	234 sq ft	n/a	n/a
Cylinders	(2) inside 19½" x 26" ins	(2) inside 18" x 26" ins	(2) inside 18 ½" x 26 ins
Grate Area	17.6 sq ft	16⅔ sq ft	17 sq ft

The SECR 'J' class tank engines

The SECR 'J' class of 0-6-4T engines - not to be confused with the J1 and J2 LBSCR types - were a small batch of just five engines built under the auspices of H Wainwright in the last months of 1913. All were constructed at Ashford, the 0-6-4T type not the most common wheel arrangement although it may be of interest to mention which other railways in the UK possessed engines of this design. Nine 0-6-4Ts were built by Beyer Peacock for the Mersey Railway in 1886, Swindon had three as crane tanks, basically an '850' class 0-6-0 with a bogie extension at the rear to support the crane, these were constructed at Swindon in 1901, then in 1904 the firm of Kitson of Leeds provided nine 0-6-4T locos to the Lancashire, Derbyshire and East Coast Railway. In time these nine became part of the Great Central fleet and in turn were absorbed into the LNER. The Midland Railway also had 40 0-6-4T engines built in 1907, these perhaps the best known engine of this wheel arrangement, possibly because of their numbers and perhaps also because of their nickname 'Flatirons'. The Highland Railway also had eight engines, four from the North British Locomotive Company in 1909 and a second batch of four in 1911. Finally were four tank engines supplied by the Yorkshire Engine Company to the Metropolitan Railway in 1915. Just one engine from the total of 78 0-6-4T engines survives, not an SECR example but instead Mersey Railway No 5 *Cecil Raikes*.

To return to the SECR however, the J class 0-6-4T engines came about, not because of the size and weight of trains operating on the line, but more so the number of trains. This last sentence could be read as referring to the whole system but that is not the case and instead it should be seen as meaning the intense service running in the London area; still steam hauled at the time in question and where existing steam stock was generally performing well, although it was recognised just one train running late could affect countless others. Locomotives on these duties were the 4-4-0 types and also the 'C' and 'H' class engines. In all cases, provided the crews did not attempt to

SR No 1595 (the original No 207). We are not told how the class rode smokebox first but a lot would have to do with the state of the track and the setting of the engine springs. Certainly we know the Midland Railway 'Flatirons' were prone to complaints of bad riding. One source that might have been expected to report on the type is within Holcroft, but he fails to provide any comment on the class. *Transport Treasury*

SOUTHERN TIMES

The 'J' was intended to be an improvement on the 'H' (left) with the original proposal incorporating the boiler from a 'C' (right). In so far as any improvement was concerned, an 'H' in good condition and when driven and fired to best effect could equal the performance of the 'J' and in addition was more free running. The 'C' had been intended to form the basis of the new design but when calculated the weight came out as too heavy. *Transport Treasury*

exceed the respective engine's capabilities, the service operated without difficulty albeit with little in reserve. Of the three designs, the 'H' class were probably the best performers and not just on these suburban services. In recent years we might recall these 0-4-4T engines as being principally branch line engines but in the years prior to WW1 they could be found on both semi-fast and cross-country passenger workings, certain turns hampering them somewhat due to limited coal and water capacity.

An obvious solution was to enlarge the design and Ashford initially came up with an 0-4-4T having a greater water capacity of 1,600 gallons, compared with the 1,350 of the 'H'. A greater reserve of power was also provided for by utilising the boiler and cylinders of the 'C' class 0-6-0 tender engines, the latter a half inch increase in diameter. Unfortunately in the process the overall weight increased to 60 tons of which 19 tons was carried by each set of driving wheels, this alone would have restricted availability whilst an additional problem was that it proved impossible to increase the water capacity above 1,300 gallons. Had the design been approved, it would have had the designation the 'K' class.

It was a case of back to the drawing board and this time the starting point were six coupled wheels with two alternative designs. The first was a large 4-6-2T and the second an 0-6-4T. The first mentioned fell foul of the Chief Engineer but he was prepared to accept the second. As a result authority was given for five engines in July 1911 at a cost of £2,790 each. Known as the 'J' class their running numbers simply took vacant spaces in the existing loco register and became Nos. 129, 207, 597, 611 and 614. This anomaly persisted until 1927/28 and Southern Railway days when the five were grouped together from 595 to 599., No 597 retaining its original number. All five were completed and in service between October and December 1913.

Initially all five were based at Bricklayers Arms working semi-fast services to Tonbridge, Hastings and Dorking Town. It might have been expected that their size would show advantages in every respect and while the six coupled wheels displayed an improvement on up gradients, particularly so on the Hastings line, they were inferior steamers to the 'H' class and failed to show any improvement with their speed capability. Indeed, referring to Bradley, 'Most men considered that a well maintained and resolutely driven 'H' was at least their equal.

Five more of the type had been ordered from Ashford in August 1913, this even before the first of the original order had been delivered, but the order was never fulfilled and was cancelled. Wainwright was asked to retire on 30 November 1913, having been partly made the scapegoat of failings which were hardly his fault. The directors insisted Longhedge Works be closed before full capacity was available at Ashford. Indeed had it not been for severe weight restrictions inherited on the Chatham lines, there might well have been Wainwright designed 4-6-0 passenger and 0-8-0 goods engines in service.

Under the stewardship of the new CME, Richard Maunsell, the class lost some of their Tonbridge work to have this replaced with Canon Street - Redhill - Reading, and Cannon Street - Oxted - Tonbridge workings, also Cannon Street to Addiscombe Road. It was on one of these services, in the Up direction, that No 614 came off the line due to a broken rail whilst approaching Ladywell. Fortunately the engine was slowing down for the stop and only minor injuries resulted. A fatality occurred at the same spot on 19 February 1915 when a trespasser attempting to evade Police arrest took a short cut across the line just as No 597 was approaching. Neither occurrence would appear to have resulted in a Board of Trade Enquiry.

Mention was made earlier of how, had the original enlarged design come to fruition in 1910, this would have been known as the 'K' class. In the event this letter was used to identify the 2-6-4T design (later known as the 'River' tanks) then being prepared for assembly. Accordingly a 'K' class bogie intended for the later No 790 was fitted to 'J' No 129 on 11 November 1916 for assessment and trial working. Initially 'run in' on local workings No 129 soon graduated to a semi fast duty normally performed by an 'L' class 4-4-0. Between November 1916 and 16 March 1917 the engine ran 14,697 miles without incident. It remained in place until the original was refitted in November 1917. (This is slightly contradictory as again according to Bradley, No 790 was completed and steamed from Ashford in June / July 1917. It certainly would not have run without a rear bogie! Likely then the dates are slightly confused.)

Motive power shortages and associated arrears of maintenance became acute following the end of WW1 and it was necessary to again use the class on the Hastings express workings including the business trains. Later, three Nos 207, 597 and 611 were transferred to Reading to work through services on and off the GWR. Once more they were perhaps not the best choice but continued this work until replaced by 'River' tanks in the Autumn of 1926.

Mention has been made of the backlog of repairs at Ashford that accrued during WW1. This affected repairs to the class as until December 1920 there was no spare boiler for the class, this despite one having been approved more than five years previously in October 1915.

Externally changes took place from December 1923 with all five starting to receive Southern green from that date;

No	Painted Green
129	December 1923
207	February 1925
597	December 1924
611	September 1924
614	October 1924

'A' (Ashford) prefixes were also likely applied around this time.

No 1599 at Archcliff Junction, Dover. Apart from the minor modifications mentioned in the text, by BR days tail rods had been suppressed on all but No 31595 while all had lost their snifting valves. In his unpublished autobiography, the late John Click specifically mentions his liking for the class but without giving reason. Possibly this may have had something to do with the fact Bulleid at one time was considering an 0-6-4T design but rejected it on the basis of weight and adhesion. That design would eventually manifest itself as the 'Leader'. *Transport Treasury*

This page and opposite, BR service including Ashford. *Transport Treasury*

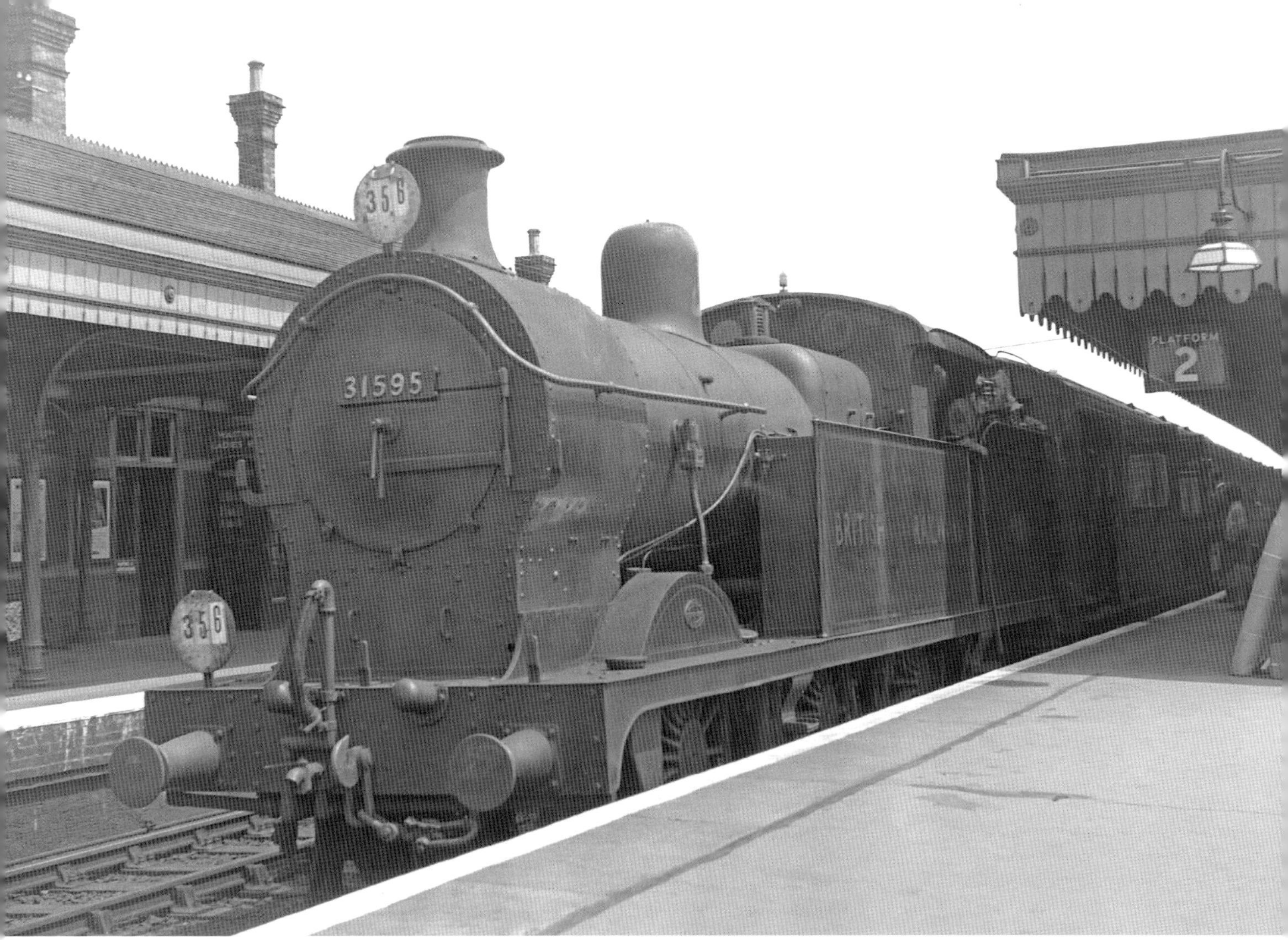
356
31595
356
PLATFORM
2

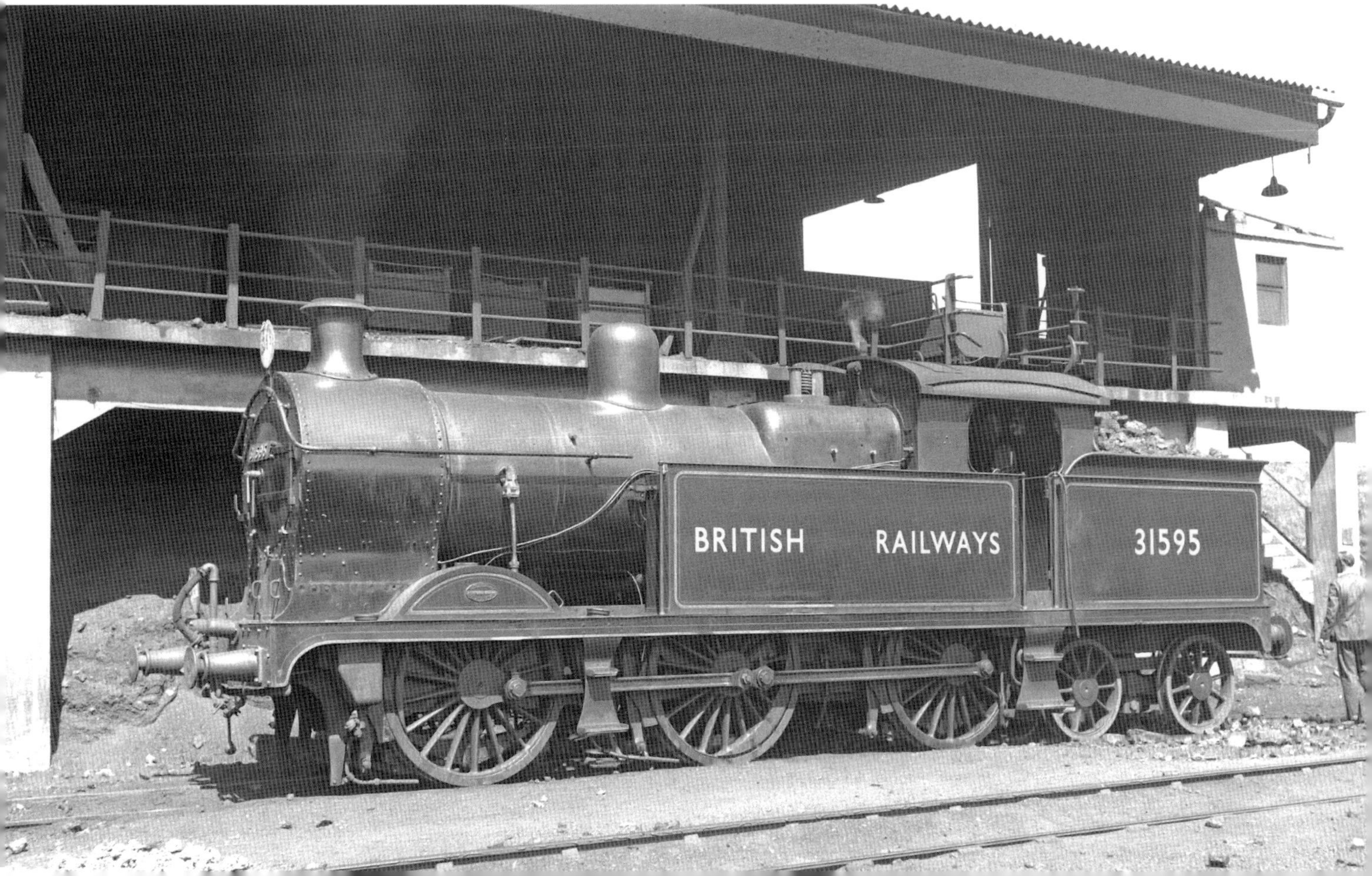
BRITISH
RAILWAYS
31595

SOUTHERN TIMES

Renumbering
A129	to A596	May 1927
A207	to A595	March 1928
A597	No change	
A611	to A598	September 1927
A614	to A 599	October 1924

Southern Railway days also witnessed a number of other external changes, chimneys fitted without capuchons, a different design of snifting valve and altered lubricators.

For most of their working lives they remained strictly Eastern Section locomotives and continued on services to Cannon Street and London Bridge. Post electrification in 1935 they moved from Hither Green, where they had been based since that shed was built, to Ashford working duties shared with the smaller 'H' class. From Ashford one of the type might also be seen on a regular Ashford to Eastbourne or Faversham to Brighton duty. This continued until 1937 after which time they were prohibited from the Ashford to Hastings line.

The class continued to be based at Ashford in WW2 working to Margate but also to be seen at Canterbury and Dover. Two, Nos 1595 and 1597 were loaned to Tonbridge during the time of the Dunkirk evacuation and found use piloting trains between Tonbridge and Orpington.

Repaints in WW2 saw all five running in plain black albeit with Bulleid 'sunshine' lettering.

All five entered BR service being renumbered 31595 to 31599. For three of the class livery was plain black but Nos 31595/96 were outshopped in lined black in March 1948 and November 1949 respectively; these were also destined to be the last members of the class in traffic. Coupled to the cleanliness possible from Ashford shed, they posed as attractive machines.

Being a small class it was inevitable their days would be numbered, this heightened with the influx of LMS design 2-6-4T locos, meaning No 31599 was the first to go in November 1949. The remainder had ceased to work by the end of September 1951.

	Withdrawn	Final Mileage
31595	April 1951	1,086,921
31596	September 1951	1,086,921
31597	October 1950	1,066,773
31598	December 1950	1,084,988
31599	November 1949	1,059,616

Reference; 'The Locomotive History of the SECR' by D L Bradley. Published by the Railway Correspondence and Travel Society.1980.

No 31596 in official pose - but withdrawn. For reasons that are unknown, No 31596 was especially cleaned after withdrawal with a series of official images taken. Immediately after it was consigned for scrap. The painted lamp irons, coupling rods etc are no doubt to effect the best contrast. *Transport Treasury*

Next time: Mr Bulleid's main line diesels, 10201 - 10203

Deep in the Test Valley between Stockbridge and Horsebridge, a 3-car Hampshire DEMU plies its trade in the early 1960s. Today much of the route is a footpath 'The Test Way', the river from which it takes its name seen on its meandering course in the background. *Roger Holmes*

The track layout at Barnstaple Junction remained deceptively impressive into the early 1970s. To the right is the former main line to Barnstaple Town and on to Ilfracombe seen in its final months of operation. Following the inevitable run down and consequent diminishing assets it would close from 5 October 1970. To the left is the subject of our journey, the original main line - that accolade had been passed to the Ilfracombe line when it had opened - running via Fremington, Instow, and Bideford to Torrington. This had closed to passengers in 1965 although occasional passenger and freight workings continued for a further 17 years until 1982.

By Withered Arm from Barnstaple Junction to Halwill Junction via Torrington

Above: The first station out of Barnstaple was Fremington and nestling on the lower tidal reaches of the River Taw. This was where ball clay from the mines at Peters Marland and Meeth was transhipped from rail to coastal vessels. The quay was adjacent to the Up platform which was noted for its elevated signal box. Clay exports ceased here from 1969 due to the physical deterioration of the dockside cranes and the quay closed in March 1970. Thereafter ball clay shipments followed a circuitous routing to reach Fowey docks.

Right: Not quite a branch train but in fact the Torrington portion of the 'Atlantic Coast Express' seen behind M7 No 30250 in July 1955.

Top: The same engine we saw on the previous page, M7 No 30250 but this time at the next station, Instow, with a relatively heavy mixed working. At the rear of the train are empty milk tanks being returned to Torrington. The passenger stock consists of WR vehicles; perhaps the load involved is the reason the fireman is feeding the grate around departure time.

Bottom: Looking east from Instow station. Following closure in 1965 the signal box was preserved and is now a landmark on the Tarka Trail. The station buildings are the headquarters of the local yacht club; all a far cry from the 1950s when the Torrington section of the 'Atlantic Coast Express' stopped here and the goods yard was flourishing. Both 21 July 1955.

Top: As appropriate to the most important town on the line, Bideford was provided with more impressive station facilities. Located adjacent to the medieval bridge across the River Torridge, it was close to the terminus of the independent standard gauge Bideford, Westward Ho and Appledore Railway which only had a short working life from 1901 to 1917 and curiously never had a connection with the LSWR. To the east, single track continued to Instow.

Bottom: Two coaches forming the 'ACE' again en-route to Torrington. Once more the same engine as before and possibly also the exact same pair of vehicles seen earlier on 21 July 1955. Just this side of the board crossing, the rails set at right angles are for a permanent way trolley.

With only a few months remaining before the end of passenger services on the line from Barnstaple Junction, there is nonetheless plenty of activity at Torrington station. An Ivatt Class 2 2-6-2T - replacement for the M7 class - is taking water after arrival with the branch train. Note also that icon of West Country bus services a Bedford OB, providing the shuttle service to the town, waiting in the station yard. In the opposite platform to the steam train, a North British Type 2 diesel-hydraulic waits to depart from the Up platform whilst overlooking the scene is a row of the ubiquitous railway cottages.

Top: A mixed train from Halwill Junction just arrived at Torrington behind an 'E1' 0-6-2T. The thin sprinkling of passengers awaiting the next service for Barnstaple attest to the inconvenient location of the station some two miles from the town. In the Southern Railway era Torrington was superseded by Ilfracombe as the principle North Devon railhead for passengers, although it retained its two coach section of the 'Atlantic Coast Express'. In the final years preceding closure to passengers in 1965 and freight in 1978, Torrington experienced a short lived expansion in bulk milk traffic and the survival of its goods shed as a fertilizer depot. This last named traffic ceased in 1980 with the final goods being ball clay handled up to 1982. Post closure the station building was reincarnated as the 'Puffing Billy' pub serving hikers and cyclists using the 'Tarka Trail' which replaced former railway rights of way between Braunton (on the Ilfracombe line) and Meeth (south of Torrington on the way to Halwill Junction - which station we shall see later).

Bottom: Like so many railway backwaters, there were periods of intense activity followed by long periods of calm. On this occasion, recorded on 20 July 1955, the down platform has just been vacated by an 'M7' tank engine and train and which in turn will form the next service back to Barnstaple. Meanwhile the stock for the morning mixed train to Halwill Junction waits more in hope than certainty for custom. The motive power for this service will be an 'E1', No 32608 which is taking water just beyond the station.

SOUTHERN TIMES

Following on from the Exeter Central recollections of John Bradbeer (ST8), John has kindly taken the trouble to write again. The correspondence between us included my mentioning that a feature on Torrington to Halwill was scheduled to appear in ST9 which brought forth a raft of welcome recollections and local knowledge, not just on the line in question but also on others parts of the SR at Exeter and westward.

We are delighted to include these starting with a recollection at Exeter:

Just a comment about the three Z class tanks in the banker siding at St David's station as shown on the rear cover. (ST8) Exeter Central was in the habit of sending bankers back to St David's in multiple, but usually in pairs. My guess is that a Meldon stone train is due fairly shortly and these always had two bankers, unlike the passenger trains, which had just the one. I attach a not very good picture of 30955 taken by my 15-year old self in 1962 banking a stone train and just about visible on the extreme left is the buffer beam of a second Z. However. my abiding memory of the bank came from the following year. Blue Circle had opened a cement depot at Central station and cement trains came down the WR main line from Westbury to St Davids and then up the incline to Central. I am afraid that I don't have details of the numbers of the locos, but the train engine was a 9F and it was given a pilot in the form of a class 4 2-6-4T and two W class as bankers, a combined power force of 26! The train literally shook the ground and I was covered by a rain of cinders as it powered up the bank. There was a brief reduction in the noise as the two front locos went into the tunnel and then I could clearly hear when they emerged at the other end of the tunnel. I had assumed that the reason that the Blue Circle cement trains came the way that they did was simply to avoid a reversal, which trains coming from Westbury would need to do at Salisbury to head on to Exeter. That was a September evening, of which I am fairly certain, as I had just started sixth form study and had gone to Exeter to buy some Pelican history books to supplement my school textbook. We had a simply awful history master, a stark contrast with my geography and economics masters, who were superb. Geoff Smith, the geographer, had grown up at Corsham on the GWR line and had seen all the Kings, which I never managed to do! Geoff was expected to coach and run school rugby and cricket teams and not to lead a Railway Club..... .

Turning now to the area around Meldon.....'I never travelled the SR on the Plymouth and North Cornwall lines that much and never ever got to Bude by train. However, Fred Cooke, my father's brother-in-law was at Meldon Junction as his first signal box after getting married. My aunt told me stories of her walking from Meldon Junction to the Quarry to catch the Saturday only train that stopped at the Quarry platform to "convey the Company's servants' wives to Okehampton" as the working timetable put it. The railway cottage is still there, but I am sure now with mains drainage, piped water and electricity, which it did not have c1930. Not surprisingly, Fred soon sought and got promotion to Halwill Junction, which had shops, a village school, a cottage hospital, where my cousins were born and a Baptist chapel, where they worshipped. In 1938, Fred moved on to Exeter Central, and so my cousins could walk to secondary school and not commute by train to Holsworthy or Okehampton. Fred felt that doing nights was well worth these advantages for the family and could never really understand why the men on the North Cornwall and Bude lines hardly ever moved to more accessible locations that entailed nights. In our enthusiasm for locomotives, rolling stock, and signalling, we tend to forget the railway wives (and children) When I tell my grandchildren of Aunt Veronica at Meldon Junction, they are pretty incredulous. Implicit in my little piece was that our family visits to Fred and Veronica in Exeter revolved around Fred's shifts (and of course my father's Saturday morning duties as a local government officer, when there was a skeleton staff in the offices and father would do one Saturday a month). So we almost invariably caught the 9.29am off Barnstaple Town and would see Fred either after early turn, or spend sometime with him before late turn.

'I don't think that my aunt ever said exactly how she got from Meldon Junction to the platform at Meldon Quarry. I presume that she (and the other signal man's wife) would have walked along the cess and down all the steps and back up again as the walkway did not cross Meldon viaduct itself. Perhaps the off-duty signalman accompanied the women and helped carry the week's shopping (in wicker baskets!)

'In the OPC reprint of the Western Division working timetable for 1932, on Saturdays, the 7.39am from Launceston stopped at Meldon Quarry platform at 8.34am and arrived at Okehampton at 8.38am. The down train for Meldon Quarry must have been the 1.00pm departure from Okehampton, calling at Meldon at 1.09pm, but unlike the up working which had an 'A' at the head of the column, there is no special indication for this train. I guess that the late turn signalman would have just had enough time to collect the women from the Quarry platform and walk back with them before signing on'.

Turning now to the North Devon and Cornwall Junction Light Railway. 'I must have first travelled that line in the summer of 1957, as a ten-year old. Even then, railways fascinated me and that summer my parents had their summer holiday at home with a Railway Runabout ticket giving access to the SR lines in Devon, which included Torrington to Halwill. My father, perhaps to satisfy his curiosity and my love of trains, brought us home from Exeter one evening via Okehampton, Halwill Junction and Torrington. I might add that our 'at home' holiday was so that we were in the house we shared with Great-Aunt Caroline over night. We did the same the following year. There was a small band of rail buffs at Barnstaple Boys Grammar School, and each year from 1960 to 1963 most of us would make one trip from Barnstaple to Torrington and on to Halwill and back on a Saturday morning. I think that we invariably had the train to ourselves. Certainly on one trip the Ivatt fairly rattled along through Hole and arrived at Hatherleigh many minutes early. The crew took on water and then adjourned to the station building to play poker. We left a minute or two late but were still on time by Petrockstow.

'In January 1960 there was a collision between a Southern National bus taking the Bideford to Okehampton route and a Torrington bound train. It seems that the train was hit by the bus as it attempted to get across the ungated crossing on the Halwill side of Meeth. The official accounts as described by John Nicholas and George Reeve in their 'Lines to Torrington' (OPC) are a bit vague, but the story I heard from my friend David, whose father was a fitter at Barnstaple Junction shed, was that the bus driver knew the train was coming and was convinced he could get across before it arrived. I guess that this was the gossip among the bus and train fitters and never given as evidence! Certainly on all my trips over the line, the loco hooted more or less continuously on the approach to Meeth from Halwill and no-one could claim that they were unaware of its approach. The bus, I think was a write off, being rather on the old side, but the loco had a dent to the front footplate steps and nothing more. I am sure, even at low speeds, that if the train had hit the bus and not the other way round then the bus would have been driven sideways and perhaps the driver of the bus injured.

'I have to say that Halwill Junction is pretty unrecognisable now from how it was back in the 1960s, although the railway cottages are still there, as is the Baptist chapel (and given how many non-conformist chapels in Devon have closed and been turned into private houses is worth remarking upon). We looked at the former station master's house in Halwill Junction when house-hunting, but it was a bit hemmed in. Of course, the original station on the Holsworthy line was just like those at Dunsland Cross and Ashbury and so when it became a junction, the station master's quarters became part of the station offices and a new house built for the station master on the other side of the level crossing.

'I might also mention the impact on Holsworthy that the railway had, or more accurately the impact that the long pause in building the line to Bude had. Holsworthy is one of those places in Devon that are a bit more than a village, but not really much of a town and the larger places nearby, especially Okehampton are not much better (W G Hoskins in his book on Devon is most disparaging about Okehampton - one street to walk up and down in ten minutes and then you're done!). However, if you look around the square in Holsworthy and at Fore Street, then you are struck by how much of the urban fabric dates from c 1870-1890, the period when the railway went no further. Yes, this is the professional academic geographer in me at work!

'On more obvious railway matters, I remember the loco shed in use at Torrington, but by the 1960s it had closed and so the 8.00am passenger train from Barnstaple was double headed to bring a loco to Torrington in time for the 8.52am departure to Halwill Junction. Similarly, the loco from the 6.30pm from Halwill Junction double headed the 8.10pm back to Barnstaple. You will probably have noticed from all the photos that almost all the tank engines based at Barnstaple Junction ran facing Ilfracombe and Torrington. The station and yard pilot at Barnstaple would have done its shunting in that direction and obviously this was far better for the loco crews. The summer Saturday exodus of Barnstaple's tank engines to bank trains over the summit at Mortehoe required that the engine or engines sent to Ilfracombe, would need to be turned there, but the pair at Braunton were ready for duty on arrival. By contrast, almost all the Bulleid Pacifics photographed on Barnstaple Junction shed would be facing Exeter having been turned at Ilfracombe on the only turntable available that was large enough. The occasional forays of Bulleid Pacifics down the Torrington line would always see them running tender first to Torrington

'At Halwill, the NDCJLR trains never really entered the original station but had their own bay on the north side of the line. There was a run-round loop controlled by a ground frame that also operated the crossover to the Bude and North Cornwall lines. The OPC reprint of the working timetable notes that there were Sunday excursions over the line linking Bude and Ilfracombe, so the shunting to move the coaches to and from the NDCJR must have been impressive. I have no recollections from my handful of visits in the 1960s that freight ever arrived at Halwill from Torrington, but there are a few pictures around showing the odd covered van that did make it there.

'As for milk trains at Torrington, Messrs Nicholas and Reeve give full details and again there are lots of shots of Torrington station showing the milk loading pipes next to the goods shed and linked to the creamery about just over a mile away by underground pipes occupying part of what had been the Rolle Canal. Remember that the Torrington extension of the LSWR largely ran on the alignment of the canal, which continued at the foot of Torrington Common to Orford Mill next to RHS Garden at Rosemoor. When passenger services ceased, the main platform, technically the up platform, was equipped with the loading pipes and the goods shed and its siding were used for fertiliser wagons for a few years. My recollections of the 1970s and 1980s at Torrington are a bit thin as I made only occasional visits, having started work at what later became the University of Portsmouth. In later years my grandchildren have now enjoyed the walk from Torrington to Watergate or from Yard to Dunsbear on the track bed of the NDCJLR.

'The clay workings at Meeth have now become a super nature reserve for the Devon Wildlife Trust (the lakes have a good range of waterfowl in winter and the wood white butterfly is as common there as anywhere in Devon). I met the former manager of the clay works on a DWT organised trip, and he told us the ECC (later Imerys) could not wait to get the railway line closed so that they could excavate the clay right up to the perimeter of the land they owned, which meant digging out the track bed. Most of the old line is now a cycle way, part of the Tarka Trail, but around Meeth new rights of way have needed to be opened up to avoid the sections quarried out in the last half dozen years of clay extraction.

'Again referring to the working timetable, there are some snippets that I have never seen mentioned in the 'standard works' on the NDCJLR. One is the maximum loads allowed by different types of locomotive Locomotives of the 'E216' class, (O2's), were only permitted between Torrington and the Marland Clay siding, and could haul just 10 wagons out of Torrington. The E1/R class is referred to as locomotives of class 'B94' and could handle 16 wagons from Torrington. The surprise for me was to see reference to class 'C757', the former PDSWJR '0-6-2Ts likewise allowed 16 wagons from Torrington. 'Class 395' is presumably the Adams 0395 goods locos, could take 15 wagons from Torrington. By 1932 the old Adams 4-4-0s which improbably worked the line at the outset had been replaced by the E1/Rs and the working timetable seems to imply that they would never be needed again by not giving their loading limits.

'Although the two running lines at Torrington converged at the road over bridge to the south, locos never used this turnout to run round trains arriving from Barnstaple, but propelled their coaches into the yard to run round there, or sometimes had the coaches taken there by another loco on duty at Torrington.

'In BR days not all the clay from Marland and Meeth was shipped from the wharf at Fremington, but some was sent to Teignmouth. I guess that the wagons were simply made up into a train at Meeth and this then picked up the Marland wagons; so those for Fremington and those for Teignmouth would only have been sorted out at Fremington, where there were many sidings and the clay trains given plenty of time for shunting, or could even have been shunted later by one of the locos on a duty which involved running light from Barnstaple Junction to Fremington. In the 1950s Fremington Quay handled more cargo than the quays in Barnstaple and Bideford.'

Opposite top: No 32608 makes a brief stop at Watergate Halt for, it was said, the sole benefit of the photographer. Passenger facilities here were almost non-existent consisting of the platform but not even any form of shelter. Behind the camera is the viaduct which replaced the original wooden structure of its predecessor, the three-foot gauge Marland Light Railway which dated from 1880. The train seen is the morning mixed service from Halwill Junction to Torrington.

Opposite bottom: At the next stop facilities (or the lack of) were the same. Nearby were a row of workers cottages associated with the ball clay industry. 21 July 1955.

Top: The North Devon & Cornwall Junction Light Railway was one of the shorter lived (1925 - 1980), least known and certainly most poorly patronised backwaters of the Southern, a single coach usually sufficient for the passenger service. As the two advertised return trips covering the length of the line were designated as 'mixed', there was always time enough for any passenger to admire the neat fieldstone stations at Petrockstow, Hatherleigh and Hole. Devon & Cornwall Farmers Ltd was a freight customer at Petrockstow where, in this view, in a less than demanding schedule there was an opportunity for the time honoured banter between train crew and station staff that epitomised the friendly ambiance of the lost age of the light railway.

Bottom: Busy times at Petrockstow with 'E1R' No 32610 on a down freight crossing classmate No 32608 on an Up Torrington train. Apart from duties on the light railway, engines of this class were to be found on banking duties between Exeter St Davids and Exeter Central. Ten engines of the type existed, all rebuilt from the former LBSCR 'E' class 0-6-0T engines. Withdrawals started in 1955 and the 'E1R' class was extinct four years later.

Unlikely power for the light railway during the final era of passenger service. This is a North British built Type 2 diesel hydraulic (number not recorded), about to leave Petrockstow with the 7.55am short working for Torrington on a misty morning in August 1964.

SOUTHERN TIMES

Top: Compared to some of the other halts on the line, Meeth did at least provide a shelter and a bench, even if the latter was outside. The arrival of the morning train was an opportunity for the usual gossip. Perhaps the fireman is pondering why anyone would ever want to take a photograph of what was then an everyday scene. 21 July 1955.

Bottom: With evidence of at least one passenger on board, No 41216 tackles the gradient between Meeth Halt and Hatherleigh with the morning Torrington to Halwill train consisting of a single Bulleid brake composite. August 1964.

Top: After the halt at Meeth the next stopping place was at Hatherleigh, the village of the same name having a population of little more than 1,000 souls but even so was the largest centre of population west of Torrington. Unfortunately the railway here suffered the same disadvantage as at Torrington with the station two miles from the town. The shortage of potential passengers was offset by a reasonable business in livestock, farm supplies and parcels.

Bottom: No 32608 pauses awhile to take water on its meandering journey between Halwill Junction and Torrington. The open air ground frame controlled access to the passing loop and was typical of the basic infrastructure on the line.

With no apologies for having used this view elsewhere in the past; but it was several years ago. An Adams 4-4-0 at Hatherleigh on the day after formal opening, 28 July 1925. *Corbis Images*

The comings and goings between Hatherleigh and Hole.

Top: Ivatt Class 2 No 41216 slows to 5mph for one of the many ungated level crossings between Hatherleigh and Hole. The service is the morning mixed train to Halwill Junction consisting of a Bulleid brake composite, three loaded ball clay wagons and a brake van. August 1964.

Bottom: From the opposite end, we see the same train as it has passed. The photographer recorded that there were two enthusiasts travelling in the brake van, adding the comment, '...an exceptional passenger load.' Not withstanding the limited revenue stream generated, the permanent way and cess will be seen to be beautifully maintained.

The final station before reaching the junction at Halwill. This is Hole (for Black Torrington), a railway posting here not the most demanding job for a member of staff. Indeed his existence would hardly be disturbed by the arrival of No 32608 on 21 July 1955. The solitary platform lamp seems to exemplify the loneliness of the location named for nothing more significant than a local farm.

Top: The light railway arrived at Halwill Junction by running parallel with the line from Bude for its final mile. At the station itself a connection with the rest of the system was made by means of a backshunt. Here we have a modern steam engine in the form of an Ivatt Class 2 attached to an item of rolling stock from a previous era; an LSWR brake coach.

Bottom: The more typical period train of 'E1R' No 32608, and single LSWR coach in the bay at Halwill and prior to running round ready for the return working. Because of the frequent need to operate mixed trains, pull-push operation was not in use on the line.

SOUTHERN TIMES

A short bare platform and footpath connected the Torrington platform to the main station, almost as if to emphasize the lowly status of the light railway once it had reached Halwill. No 41295 hopefully waits for a rare and adventurous sole who might arrive on the next train while a connecting service to Okehampton and other points east is ready to depart from the Up platform. Halwill Junction - the name was indeed appropriate at what was a railway crossroads, saw services west to and from Bude and Padstow and in the opposite direction Okehampton and Exeter. There was of course another way to travel by train between Halwill and Torrington which was to go almost in a circle east to Okehampton and on as far as Yeoford Junction between changing directions to reach Barnstaple and then finally south back to Torrington.

Next time: Demolition at Crystal Palace by R C Riley

Farnborough Air Show Traffic
From the notes of John Davenport

The Southern Railway and its predecessors were well versed in the business of moving large numbers of passengers to and from sporting events and other similar gatherings. This could even be said to have begun with the 'siege of Nine Elms' on Derby Day 1838.[1]

Move forward a century and by the end of the 1940s, the annual Exhibition and Flying Display of the Society of the British Aircraft Constructors had become highly popular. The venue had also been changed from Hendon to the Royal Aeronautical Establishment (the R.A.E.) at Farnborough because of space constraints both on the ground and in the air; the event became popularly known as 'The Farnborough Air Display'.

The move enabled the catchment area for public visitors to grow and the space for them to be enlarged. It became traditional for the Exhibition to be held in the first week of September each year - Monday for rehearsals, Tuesday to Thursday for serious visitors and dignitaries, Friday for higher-priced public entry and Saturday and Sunday for the general public. In 1952, the last three days were 5, 6, and 7 September.

According to 'Flight' magazine in its issue of 12 September 1952, there were some 300,000 members of the public in attendance over the three days - 40,000 on Friday, 120,000 on Saturday and 140,000 on Sunday. While not all of them travelled by train, car ownership was not yet widespread and they and visiting coaches had to cope with an unimproved road system. As a rough comparison, some 338,000 personnel were carried by the railways during the Dunkirk evacuation over a greater number of days between 27 May and 4 June 1940, as recorded by Peter Tatlow in 'Return from Dunkirk', Oakwood Press (2010).

A point of interest is that the contemporary popular edition of the Ordnance Survey map covering the area 'fully revised in 1930 with later amendments to 1947' omitted the R.A.E. completely on the grounds of national security. The R.A.E. had in fact been established in 1914 and so very early on in the history of aviation. To those in the know its position might be determined from the single track railway siding running southwest from Farnborough Main station. Road access was from the A325, which ran northeast - southwest between the A30 and A31. As to rail routes serving the location there was the South Western main line at Farnborough Main, the Alton branch at Aldershot, and the Reading South - Redhill via Guildford line at North Camp. The Guildford - Ascot line provided additional services to and from Aldershot. [2]

For the event the Southern Region London West District produced a 43 page Supplementary Notice of Special Working Arrangements for the three days. Full of detail, it gave opening times for signal boxes such as Farnham Junction, Aldershot Government Sidings and Foxhills on the Alton line, and Farnborough North, Blackwater and Crowthorne on the Reading line. There was also a modification to Block working (Regulation 4) at Aldershot east signal box with the signal box itself specified as the temporary clearing point. This allowed more trains to be handled and in a quicker time. Special instructions were also given for specific members of staff to be available at set points.

1. From 'The Times' and 'Consuming Passions: Leisure and Pleasure in Victorian Britain' by Judith Flanders, published by Harper Press (2006), 'Nine Elms Station, the London terminus of the London and Southampton Railway was opened in 1838 .... just nine days after the opening, eight special trains ran to Epsom and carried more than 5,000 people for the Derby.

'The steam-boats which ply from London-bridge (sic) and from Hungerford were filled with passengers, who made sure of getting down to Epsom by the railroad. Hundreds were fated to be disappointed. There were ten times more applicants for seats in the train-vans than there were seats for their accommodation. The proprietors did what they could to meet the demand for conveyance, but they could not do what was impossible.

'Sir John Easthope, the chairman of the London and Southampton Railway, was a keen follower of horses, and within a week of the opening of the line from London to Kingston, in 1833, the company had scheduled eight special trains to take spectators to the Derby. At Kingston, there was a long walk from the station to the racecourse, but such was the enthusiasm that, after the seventh train had left Nine Elms station in south London, 5,000 would-be spectators were still waiting to board the final train. When they realized that most of them were not going to reach Kingston, they staged a riot.'

2. Both Aldershot and North Camp stations were used to dealing with heavy military traffic, so there was space for passengers bound for the display to leave the platform and queue for the buses running the shuttle service. The Aldershot and District Traction Company was at full stretch, aided by other operators.

Unfortunately most of the images we were able to source for this piece are not necessarily from 1952. Here a down special for the 1956 (Saturday 8 September) event has just arrived at Farnborough Main behind No 30498 *Sir Hectimere* and from the look of the stock has originated on the London Midland Region. Presuming the same criteria applied as in 1952, as soon as the train had unloaded it would continue to Basingstoke for servicing. *Photos from the Fifties 94B / Transport Treasury*

In 1952 British Railways were using the 12hr clock whilst Saturday 6 September was also still in the Summer service, which limited the possibility of extra main line services.

As an example, the regular main line services to and from the West Country and Bournemouth lines, had, at Woking, to be slotted in to the existing four fast electric services an hour on and off the Portsmouth line. In addition there were stopping services, those to Portsmouth and Alton separating / joining at Woking (dependent upon direction of travel), which at least relieved a little in line occupancy. Even so the signalmen in the manual boxes from West Byfleet to Esher were kept very busy.

On the Friday there were only two amendments to ordinary working. The stock off the 7.10am Yeovil to Waterloo was to be used without detachment for the 11.54am Waterloo to Salisbury and the 5.39pm to Salisbury was not to be reduced in size at Basingstoke. Both trains called at Farnborough Main. For the rest of Friday action was confined to the Alton line with four additional down trains from Waterloo. A similar arrangement occurred at the end of the day, apart that is from some additional empty stock movements ready for the next day.

On Saturday Farnborough Main was served by special stops on regular services and by some extras. As an example between 11.12am and 11.35am three such trains called. The first was the 9.50am, from Southampton Central, then at 11.19am a service from Bournemouth Central, and at the last time mentioned a working from Exeter Central. In all such cases the Special Notice made specific mention that the Guard and Driver of each of these workings was to be reminded at the previous stop of the need to call additionally at Farnborough. The respective station masters were responsible for seeing this was carried out. A similar arrangement had applied to the trains calling earlier in the day. Interestingly the Special Notice specially refers to the additional stop at Farnborough main being to 'Set Down' or 'Take up' only. (Adjacent traffic areas had similar reminders as to the extra stop in their own weekly notice.)

The prestige extra of the day was a special from Derby for Rolls Royce Ltd, identified by the reporting number 'M980'. This was made up of ten coaches including two restaurant cars and routed via Brent Junction, Kew East Junction and Brentford; it reached the main line at Byfleet Junction at 11.50am and Farnborough Main at 12.12pm. After disgorging its passengers it ran empty to Basingstoke with the instruction, 'Clean and water train, gas and water restaurant cars. Engine headboard to be retained for the 6.10pm empty train' - the headboard probably referring to the train identification number.

For the evening return there were four special stops made by down trains. The 5.00pm to Exeter Central at 5.48pm, the 5.30pm to Bournemouth West at 6.12pm, the 6.00pm to Plymouth at 6.42pm, and the 6.30pm to Weymouth at 7.12pm. Farnborough also provided its own semi-fast train to Southampton Central at 6.00pm; this started from the Military Siding behind the down platform, calling at all stations to Basingstoke and then only at Winchester and Eastleigh.

In the up direction there was an extra at 6.30pm for Woking and Waterloo, with stock berthed in the up sidings from the 2.45pm empties from Waterloo. It was followed at 6.40pm by the return of the Rolls Royce special to Derby. This train was due at Dudding Hill Junction at 8.09pm, at which point the Southern engine was detached and left for Feltham at 8.25pm. Probably a similar engine change had occurred for the outward run, it was certainly the case in later years, but this does not appear to be mentioned.

It was the electric services on the Alton line that were to carry the majority of passengers on both Saturday and Sunday. Each special train both up and down, carried an identification letter in the offside window of the motorman's cab. On Saturday down trains were lettered 'A' to 'L' and up trains 'A' to 'O'. For Sunday the letters 'A' to 'R' and 'A' to 'T' respectively were used Special instructions stated all cab doors were to be locked and windows closed. On the down line between 8.54am and 2.02pm there were nine of the regular Surbiton and all station services to Alton at half-hourly intervals interspersed with 14 special trains. To be fair, three of these were not technically 'specials' but instead 'Saturday business trains'. In 1952 it was not uncommon for offices to work a half day on Saturday and in consequent three trains from Waterloo to Farnham left between 12.17pm and 1.17pm.

Regular and special workings at Farnborough Main on 8 September 1956. No 35001 *Channel Packet* is passing non-stop with a West of England service, whilst in the military platform S15 No 30838 waits with stock for one of the specials. The single disc headcode on the S15 indicates a Waterloo - or in this case, Farnborough, to Southampton working. *Photos from the Fifties 95A / Transport Treasury*

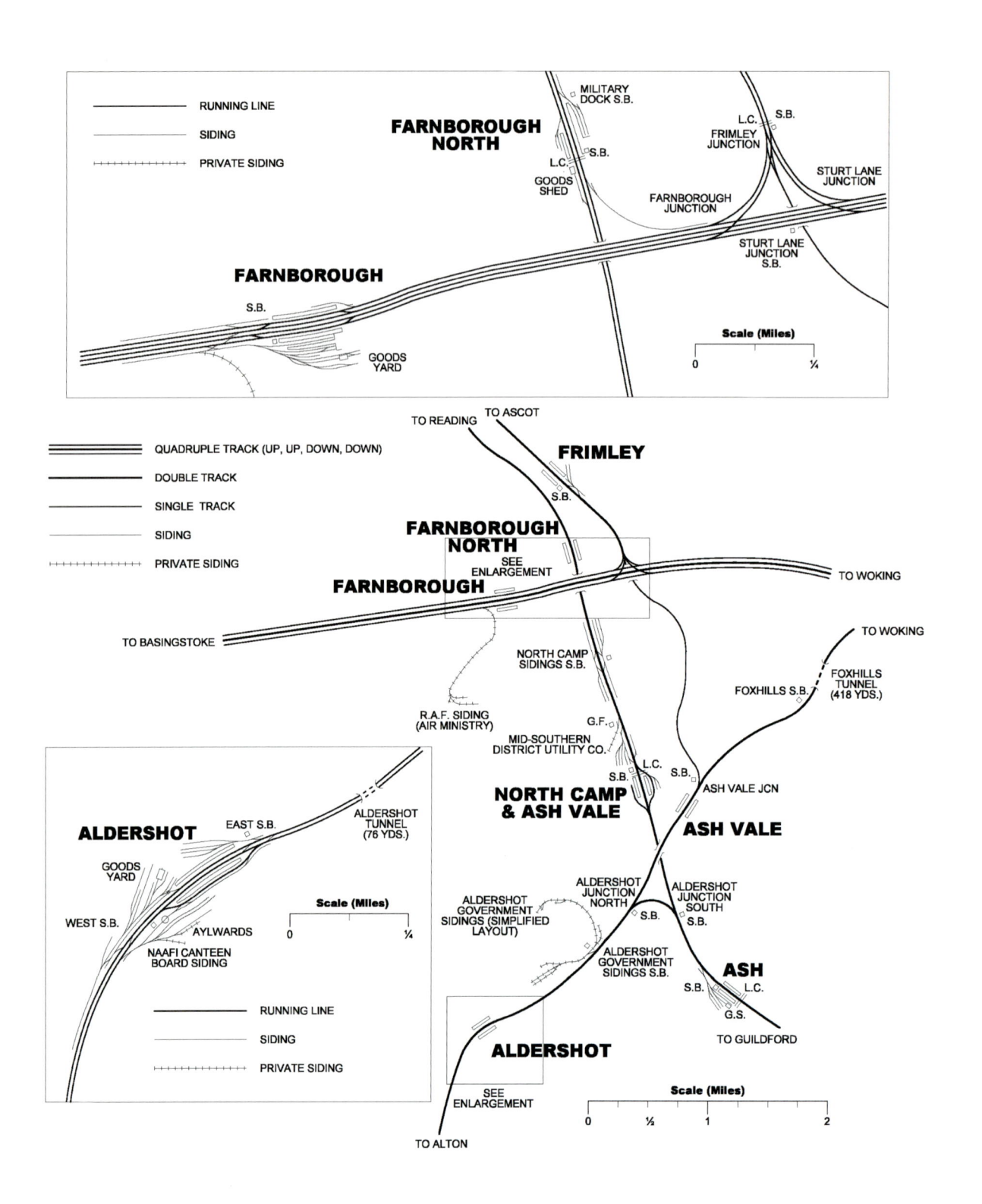

Above: Railways in the Farnborough area with locations and junctions referred to in the text shown. Coutesy Richard Harman

Opposite page: 'U' class 2-6-0 No 31631 on a special Cambridge to North Camp working approaching Farnborough North. The headcode was carried over from when the train had arrived on Southern Region metals. *Photos from the Fifties 8284 / Transport Treasury*

Under normal working arrangements, the regular Alton service consisted of either a two or four coach train dependent upon the time of day; as mentioned often splitting / joining a Portsmouth service at Woking. For the special weekend it is likely the Alton trains were made up of eight cars or even 12 if running as a special from Waterloo. Surprisingly, formations are not mentioned in the Special Notice and were probably subject to a separate instruction. What we do know is that a spare electric train complete with motorman and guard was held in reserve at Farnham throughout.

Another point is that apart from John Davenport's notes, there appears to be no mention of what was a considerable increase in traffic in either the 'Railway Magazine' or 'Railway Observer'.

Readers will already have gathered that a few extra trains and additional stops at Farnborough Main could not hope to have coped with the number of visitors quoted and instead we now turn to other locations from where visitors were bussed to the display. Principal amongst these was Aldershot, served by Alton line trains. During the morning of the Saturday, busy would be an understatement but it was nothing compared with what would occur for the return with 20 trains despatched between 5.09pm and 8.00pm. There were just two extra trains on the Reading - Guildford line.

In overall charge of the railway services were the control staff at Woking

September 1952 will likely not be remembered for the additional train services to Farnborough but instead more for the tragic accident that occurred over the airfield during the afternoon. At 3.45pm the prototype De Havilland 110 twin engine fighter broke up during its display. The crew of two died as did 29 members of the public as the debris fell on to one of the areas where people were stood. A further 60 were injured. The disaster was described in 'Flight' magazine the following week who reported, 'Demonstrations continued until 3.50pm when the DH110, having made a supersonic arrival from Hatfield, broke up while flying over the airfield..... .the flying programme continued. At the time, the DH 'Venom' was waiting to take off and it did as soon as the runway was clear. Police and medical personnel were assisted by the Bristol 171 helicopter...and by the flying display itself which held the attention of the main part of the crowd.'

The same magazine later commented that perhaps the majority of those attending were still influenced by wartime disasters in the acceptance of what had happened together with the unemotional tone of the reporting. As an aside, the accident did lead to instructions that remain in place today, namely that

SOUTHERN TIMES

display aircraft shall not perform directly over crowds.

Notwithstanding the tragedy, the Sunday display and consequent special trains went ahead as normal. On the main line, there were five additional stops (and later pick ups) made. There was also a special from Cambridge routed by Potton, Sandy, Bedford, Bletchley and Reading. Here the train joined SR metals (details of any motive power change are not known), after which it called at Farnborough North before reaching its destination at North Camp, again buses were used from this point. The stock returned to Reading South for servicing. Although not confirmed it is possible this trip was organised by the Royal Observer Corps.

So far as the railway was concerned, hopefully all the stock was back in its allocated space by the end of Sunday and so ready for the Monday morning commute.

This page, top: Not exactly express motive power, Q1 No 33005 on a return service from the air show heading for Reading and recorded near Farnborough North.
Photos from the Fifties 8110 / Transport Treasury

This page, bottom: Some of the crowds queued for bus services to the show on a warm September afternoon, Sunday 13 September 1959.
Photos from the Fifties 507C / Transport Treasury

Opposite top: Cardiff to North Camp special, behind another 'U', No 31627 and also near to Farnborough North. The starting point for this train is not mentioned in the STN and may simply be one of the 'additional' services referred to as starting from Reading. It is a matter of some regret that no images have been found of the electric units / trains involved in the workings.
Photos from the Fifties 8022 / Transport Treasury

Opposite bottom: A pair of Moguls, the lead engine identified as 'U' No 31798 joining the main line at Sturt Lane Junction with ecs for a return air show special. The stock is of Eastern Region origin and so this may well be the Cambridge service.
Photos from the Fifties 95A / Transport Treasury

Opposite insert: the cover of the 1952 Special Traffic Notice. We should not forget that apart from the additional workings there were any number of alterations, additions and notes, these included terms such as 'Will not run' and 'Pathway not available'.

BRITISH RAILWAYS

SOUTHERN OPERATING AREA

London West District — SPECIAL NOTICE No. P.52 L.W.D., 1952.

SUPPLEMENTARY NOTICE OF

SPECIAL WORKING ARRANGEMENTS

ON

FRIDAY 5th, SATURDAY, 6th and SUNDAY, 7th SEPTEMBER

IN CONNECTION WITH

THE SOCIETY OF BRITISH AIRCRAFT CONSTRUCTORS' FLYING DISPLAY AND EXHIBITION AT FARNBOROUGH

For General Instructions governing the working of Special Trains, see General and Western Section Appendix to Working Time Tables.

For Announcements to the Public, see pages 123-126 of the General Appendix.

Attention is directed to paragraph 61 of the Instructions applicable to Electrified lines, and particular care must be exercised by all concerned to ensure that all motormen's cab doors (both sides of train) are locked and windows closed.

EXPLANATION OF REFERENCES

A	Additional Stop.	M	Via Main line.	V	If late, stop not to be made.
B	As shown in Service Timetables or Supplements thereto.	Q	If required.	W	Via Windsor line.
EP	Via East Putney.	S	Shunt.	§	Light engine.
L	Local line.	T	Through line.	†	Empty Train.
		TW	Via Twickenham.		

22nd August 1952.

(B.48½/4743)

S. W. SMART,
Superintendent of Operation.

Stephen Townroe's Colour Archive
The 1952 Shawford derailment

Stephen Townroe in his position as Shedmaster and later as District Motive Power Superintendent at Eastleigh was also responsible for the investigation of accidents in his area. In addition he attended several of these in person: hence the archive contains a number of accident related views.

One of these occurred on Sunday 20 July 1952 immediately south of Shawford station. Fortunately, or unfortunately dependent upon one's viewpoint, the Townroe family home was also nearby and he was quickly on the scene.

The lead up to and subsequent investigation is perhaps best described by Brigadier Langley from the formal Ministry of Transport report.

"I have the honour to report .....into the derailment which occurred at about 3.58pm. on Sunday, 20 July, 1952, at Shawford, near Winchester, on the Waterloo-Southampton four track main line in the Southern Region."

The 3.24pm. Southampton Central to Waterloo passenger train, comprising seven coaches drawn by a 4-6-0 type Lord Nelson class engine (No 30854 *Howard of Effingham*), was approaching Shawford Station on the Up Local line at about 30 m.p.h. when the driver mistook the signals ahead and passed the Up Local Home at Danger. This signal, which controls the exit from the Up Local to the Up Through a quarter of a mile ahead, was being maintained at Danger in order to give preference to a Southampton Docks to Waterloo Boat Train on the Up Through line. The local train over-ran for a distance of 560 yards and went through the sand-drag at the end of the trap siding. The engine overturned down a 20 feet high embankment and was followed by the tender and leading coach, both of which were derailed, hut fortunately remained upright. There were no casualties among the 70 passengers, nor were any of the crew injured. The passengers were transferred to the Boat Train, which had been stopped by the guard of the derailed train.

The permanent way was undamaged except for one rail length in the sand-drag, and none of the adjacent lines were obstructed. The coaches of the disabled train were drawn back to Eastleigh and the Up Local line was re-opened for traffic at 6.24 p.m. the same evening, when normal working was resumed, though the overturned engine was not recovered until ten days later.

The overall length of the train was 528 feet and its total weight, including the engine, was 373 tons; the vacuum brake operated on the coupled and tender wheels of the engine and on all the wheels of the coaches. The total brake power was 230 tons, equivalent to 62% of the total weight. There was remarkably little damage to stock. The leading bogie of the first coach was forced

Looking north on the approach to Shawford station. The signals (electrically operated from Shawford Junction signal box), are the Shawford station Up home Local and Up Through line signals. The line to the left is the Up local which ends in a sand drag, just over 440 yards ahead of the post. (440 yards being the recognised 'clearing point' - basically the provision for an over-run).
All images in this article by S C Townroe

into the underframe and some of the under gear was broken and twisted, but the engine and tender received superficial damage only. It was a fine warm day with a slight breeze.

Description

The main line from Southampton to Winchester runs approximately north between Eastleigh and Shawford, which are 3¾ miles apart. As will be seen from the diagram the four-track section ends at Shawford, where the Up Local joins the Up Through, a quarter of a mile south of the station. The curvature generally is slight and the view of the signals is good. There is a flat left hand curve near the Shawford Station Up Distant signals and there is a deep cutting between them and the Home signals, after which the line runs on to a short embankment before reaching Shawford Station. The block sections are Eastleigh to Allbrook and Allbrook to Shawford Junction (¾ mile north of Shawford Station). There are intermediate block section signals at Otterbourne between Allbrook and Shawford Junction.

The facing connection leading from the Up Local to the Up Through consists of a 1 in 12 crossover with short trap siding and a 138 feet long sand-drag in continuation of the Local line. The sand-drag ends at the edge of a 20 feet embankment down which the engine overturned. The Shawford Station Up Home Local and Through signals, which control movements through the

Top: The sand drag at the end of the Up Local, reinstated after the mishap. No 30854 had continued along this to the end before rolling over to the left. Approaching on the Down Through line is a Bournemouth working - details not given.

Bottom: The engine on its side two days after the event. Apart from the first coach of set 865, other vehicles of the train have been removed, The front vehicle remains in situ due to a damaged bogie. In a different era to today, the Up Local line was restored to traffic less than 2½ hours after the accident; this short time including the sending of an engine to attach to the undamaged vehicles and with the Up Local to Up Through crossover clear, first pulling these back a short distance and then probably propelling them forward before setting back on what was then a facing connection to return to Eastleigh. The boat train passing the recalcitrant No 30854 and occupying the Through line at the time of the accident was hauled by King Arthur No 30749 *Iseult* and had 250 passengers on board. The Eastleigh breakdown crane arrived on-scene at 6.52pm, the delay due to the amount of Up traffic which was now being limited to using the Up Through line only. It took some time for the backlog of trains to be cleared. Details of how and when the single coach was rerailed, but certainly involving a possession of all but the Down Local line, are not known. Following its arrival on-scene, the Eastleigh breakdown was temporarily stabled in Shawford goods yard but with no decision yet made on how to deal with the engine, it returned to Eastleigh depot. 22 July.

SOUTHERN TIMES

Top: The tender from No 30854 was lifted clear from site during the early hours of one morning; this being the only time available when traffic flows permitted its extraction. Management hierarchy, likely SCT (or certainly him as one of the decision makers) had come to the conclusion the engine would need to be dug out and then lifted to a normal position with rails attached to its wheels. Plant was hired from the local firm of Sellwood from Chandlers Ford, their excavator seen in the early stages of what was known as 'Stage 1'. 25 July.

Bottom: Preparing the base, Stage 2', the engine is now supported on sleepers and the excavator is making a pathway to where the sand drag ends. Temporary track will then be laid. The excavator had dug to a depth of 4' 6" below the engine, the weight supported on sleepers. Sandbags were also used to temporarily support part of the excavated embankment. Meanwhile the crew take a welcome breather. In the course of recovery, brake-gear, footsteps and injectors were removed whilst by tying the wheels to the groove in the rails the flanges were guaranteed to remain within the rails. 28 July.

Opposite page: The steep area of Shawford Down overlooking the railway provided for a grandstand view of proceedings during what was a hot week. The Eastleigh breakdown train can also be seen stabled in Shawford goods yard. Apparently the location became quite a spectator spot. 26 July.

crossover and along the main line, are 440 yards in rear of the Local line facing points and are on the west side of the line between the Up Local and the cutting. The Distant signals are a further 840 yards in rear. Both sets of signals are upper quadrant semaphores, and each pair is mounted on the one main post with the two arms bracketed level with each other and 6 feet 3 inches apart. The view of both pairs of signals is good; the Distant arms can be seen at a range of 900 yards and just after passing them the Home arms come into sight 800 yards ahead. The Allbrook and Otterbourne signals are of the same type and each pair is also mounted on a single post to the left of the Local line.

The Shawford Station Up Distant and Home signals and the Local to Through crossover points are worked electrically from Shawford Junction signal box, which is just over a mile to the north of the facing points. The maximum authorised speed through the 1 in 12 crossover Up Local to Up Through is 20 m.p.h.

Report

The 3.24pm. Southampton Central train left Eastleigh on time at 3.48pm but was routed on to the Local line because the Southampton Docks to Waterloo Boat Train, which was running late, had been accepted on the Through line. All the Allbrook and Otterbourne signals, both Local and Through, were clear for the two trains but at Shawford the Up Local Distant was at Caution and the Up Local Home at Danger with the crossover set for the sand-drag, so that the Boat Train could have a clear run on the Through line.

Signalman B R Sparkes of Shawford Junction box, explained that he accepted the Boat Train on the Up Through from Allbrook at 3.45pm and, having obtained 'Line Clear' for it, he pulled off the Through line signals. Three minutes later he accepted the 3.24pm train in the Up Local, but, having cleared the line for the Boat Train he kept the Shawford Station Up Local Home at Danger and the Distant at Caution; he satisfied himself by glancing at the repeaters that the signals were actually in the 'On' position. Examination of the signalling apparatus after the accident showed that the interlocking was in order so that the Up Local Home could not be cleared when the crossover was set for the sand-drag, and the reversal of the Up Through Home signal lever locked the crossover points in that position.

W G Greenough, who was the driver of the 3.24pm train, confirmed that the Allbrook and Otterbourne signals were clear when he saw them from his position on the left side of the engine. He said that the Shawford Station Up Through Distant was 'Off' but the Up Local Distant was at Caution when he approached it; he, therefore, closed the regulator, and shortly after passing the signal he made a partial brake application. His engine was making a lot of smoke, which obscured his vision so much when he shut off steam, that at first he did not see the Home signals ahead. He caught a glimpse of a signal in the 'Off' position when he was about 100 yards from it, and promptly accepted it as applying to the Local line, though he admitted he could not see the second signal on the same post. Having assumed wrongly that the line was clear he opened the regulator again and did not realise be had over run the signal at Danger until he saw the facing points set for the siding instead of the crossover. He made a full brake application but this was too late to stop the train, which entered the sand-drag at 20 to 25 m.p.h. Greenough remained in his place on the footplate but his fireman was thrown out when the engine overturned down the embankment. Fortunately neither was hurt. Greenough said that he was based on the Nine Elms Motive Power Depot and he seldom travelled over the Eastleigh-Winchester route. He had learned the road about 10 years ago, and last April he spent a few days as an extra man on the footplate refreshing his memory. Since then he had only worked one train over this line. He felt perfectly confident about his knowledge of gradients and signals, though he said "it did baffle him a bit going up on the Local, as he had always been used to going on the Through line with a passenger train". He had no complaints to make about the signal aspects and he confirmed that his brakes were in good order. He frankly admitted his mistake, and he agreed that he was too hasty in his action and that he should

SOUTHERN TIMES

have stopped when he saw only one signal instead of two. He is 54 years of age with 32 years Railway service; he has been a driver for the last five years.

Fireman R C Manning could add little to his driver's evidence. He explained that he had small coal in the tender and had some difficulty in maintaining steam. He did not notice the aspect of the signals nor the line on which they were running after they left Eastleigh because he was busy practically the whole time attending to his fire. He did not realise anything was wrong until the driver suddenly turned round and told him to 'jump for it'.

F L Baker was the guard of the 3.24pm train and was travelling in the fifth coach from the engine. He tested the brakes before leaving Southampton Central and found them in good working order. He said that the train left Eastleigh on time and that he observed the signals through the periscope in the van. As the train came round the curve past the Shawford Distant signal at a speed of 25 to 30 m.p.h. Baker saw the Up Local Home at Danger and the Up Through Home at Clear. The train did not slacken speed and Baker said he was about to operate the brake in his van when he felt an application and noticed the brake gauge drop from 21½ inches to 19 inches. By this time the train was close to the Home signal, which was hidden by the black smoke coming from the engine; it did not stop, however, and shortly after passing the signal post Baker felt one or two bumps and realised that something unusual must have occurred. He explained that he thought the driver must have seen the Local line signal clear for him, and it was not until his van had passed the signal that he realised something was wrong, by which time it was too late for him to take action. Baker got out of the train as soon as he could and ran back with his flags and detonators to warn the driver of the Boat Train which was approaching on the Up Through line. His signal was seen in time, and this train stopped a short distance in rear of the other.

F W Hodgkin was acting as Assistant Guard of the 3.24pm train and was travelling in the brake van of the leading coach. He was sitting opposite the rear periscope and was looking back along the line. He did not notice the aspects of the Allbrook and Otterbourne signals but he thought that both the Shawford Distants were at Caution. The train was travelling at about 35 m.p.h. when it passed these signals, and Hodgkin felt the driver make a slight brake application which reduced speed by about 5 m.p.h.; the vacuum gauge dropped from 20 inches to 18 inches. The next thing Hodgkin remembered was an emergency application, which was made when the train was well past the Home signal. He agreed that the train engine was making a lot of very black smoke.

Driver S Bracher, who was in charge of the Boat Train engine, said that he was travelling at about 50-55 m.p.h. on the Through line between Eastleigh and Shawford. The Allbrook and Otterbourne Up signals, both Local and Through, were 'Off' and as he was approaching the Shawford Station Distant signals, he caught a glimpse of the other train on the Local line; the Through Distant signal was Clear, but the Local Distant was at Caution. When he came round the curve he expected to see the local train standing at the Home signal, but instead his fireman shouted to him that there was a red flag ahead. Bracher immediately made an emergency application.

Conclusion

There is no doubt that the Shawford Station Up Local Home signal was at Danger and the Through signal on the same post was at Clear when the 3.24pm train over-ran it. The responsibility for this accident therefore rests on Driver Greenough, who frankly admitted his serious mistake. Although his recent experience of this route was slight, I do not think he was confused by the signals, which stand out clearly. The smoke from his engine may have temporarily obscured the view of the Up Home signals, but when he did see an arm in the 'Off' position, he was much too hasty in jumping to the conclusion that it was for his own line and not for the Through route. The weather conditions could scarcely have been better and although smoke might have been hanging about a little in the cutting, I am confident that it could not have seriously affected Driver Greenough's view if he had been keeping a proper lookout. In any case, having seen only one signal ahead he should have been prepared to stop, and he should not have passed it until he had made certain it referred to his line.

Guard Baker appears to have kept a good lookout for

Opposite top: 'Stage 3'. Rails have been attached to the wheels and the engine is slowly being jacked upright; Bill Bishop recalls that one of the few places the hydraulic jacks could be sited that was not curved was against the firebox wash-out plugs - at least four jacks were used. It was during the early stages of the recovery an accident occurred; the tender in the course of being removed when a lump of coal became dislodged and hit one of the fitters on the head necessitating a trip to hospital and stitches. A similar recovery exercise involving jacks and rails tied to the wheels was necessary when a Battle of Britain Pacific was derailed at Hither Green a few years later.

Opposite bottom: Temporary track being laid back towards the end of the sand drag. Lifting the engine using cranes had been thought of but it was considered the embankment would not support the weight of the cranes, 30 July.

SOUTHERN TIMES

signals, which he could see through the periscope in the van of the fifth coach. It is therefore unfortunate that having seen the Shawford Up Local Home at Danger, he took no positive action to stop the train when it ran past. His view of the signal may also have been temporarily obscured and when the train did not stop after the brake application Baker may have thought the driver had seen the signal clear for him. If, however, he had continued to look out he ought to have seen that the signal was still at Danger when his van passed it and he should then have had time to make an emergency brake application. This could have stopped the train before the engine reached the overturning point on the bank, even if speed had been slightly greater than the driver's estimate of 30 m.p.h. because the engine would still have had 400 yards to run before it reached the end of the sand-drag.

Remarks

Three months before this accident, Driver Greenough had spent seven days refreshing his memory of the route, after having learned it 10 years previously, and he should have been quite familiar with it. He was, however, on an engine link whereby he only worked over this route on Sundays, twice in 24 weeks, at intervals of 10 and 14 weeks respectively. Although I do not think it was a factor in this case, I consider that these periods are too long, and do not give a driver sufficient opportunity of keeping himself familiar with conditions. This particular Sunday duty has now been transferred to another link so that Nine Elms Depot drivers will normally work over the route not less than one week in five.

Although it is the practice in modem installations to site each signal wherever possible alongside the line to which it refers, there are a number of quadruple track routes where the signal arms for two parallel adjoining lines are placed on the same post. No objection can be taken to this arrangement provided the signals stand out clearly and their indications are unmistakable, as they are at Shawford."

'Stage 4'. No 30854 near the top of the slope. It was pulled to this position using Kelbus rerailing gear - which was in effect a reduction gearing mechanism. The prime mover for this task was not as might have been expected another engine, but instead a bulldozer again hired from Messrs Sellwoods and which was noticed to have a suitable winch on the rear. The wire hawser is attached to the engine dragbox with the wires also visible at leg level. Had a locomotive used the Up Local line would have been obstructed. The final stage was to pull the locomotive back on to the running lines, the driving wheels taking this move easily although the bogie had another idea at first but which was quickly dealt with.

SOUTHERN REGION
DERAILMENT AT SHAWFORD (HANTS)
20th. JULY, 1952
(Relevant signals only, shown)

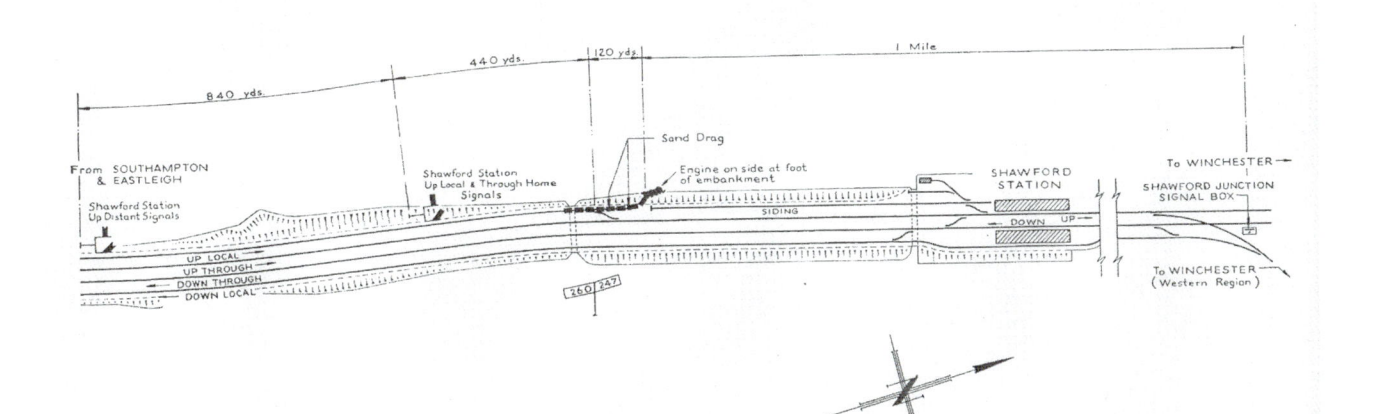

NOT TO SCALE

Two of the Townroe sisters wave at a Down train from what was known as Bowker's footbridge near to their family home at 'Collerton'. This is Shawford cutting with the Up line signals referred to in the report visible. It was a mistake in observing the Up local line signal at danger and a belief the train was running on the Up Through line that caused the accident.

Next time: Around the Southern Region
Don't forget, copies of the **SCT** colour images are available as downloads.

In and around Fratton
Images by Tony Harris

Fratton steam shed has long been the poor relation of steam depots in Hampshire. Following electrification prior to WW2, it was usurped from its position as home to locomotives working the prestige passenger services on the Portsmouth direct line but nevertheless continued in operation right until the end of steam, providing servicing and stabling for engines working parcels and freight trains, as well as being home to the 'Terriers' operating services on the nearby Hayling Island line.

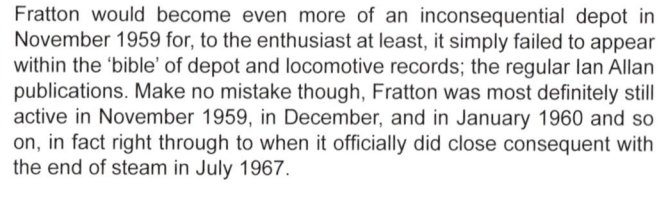

Fratton would become even more of an inconsequential depot in November 1959 for, to the enthusiast at least, it simply failed to appear within the 'bible' of depot and locomotive records; the regular Ian Allan publications. Make no mistake though, Fratton was most definitely still active in November 1959, in December, and in January 1960 and so on, in fact right through to when it officially did close consequent with the end of steam in July 1967.

As for who to blame for this omission, well perhaps that lies in the hands of the motive power department of BR who decided to change the status of Fratton from being a sub-shed of Nine Elms to a sub shed of Eastleigh. Any number of large sheds regardless of region had a number of smaller sheds within their jurisdiction. In the case of Eastleigh this was Andover Junction, Lymington, Southampton Terminus and Winchester, and should of course now have included Fratton. The fact Fratton was on paper at least no longer present under the heading for Nine Elms and somehow was omitted from the Eastleigh listing is the most likely reason for the mistaken belief it had closed. This information is taken from the notes of 'What Really Happened to Steam', penned by Roger Butcher.

In the accompanying views by Tony Harris, we have:

Opposite top: No 76033 outside the depot on 8 August 1965. The background is interesting as the building behind the loco had been provided in 1947 as part of the ill-fated oil burning episode whilst the concrete supports for the oil tanks can also be seen.

Opposite centre: Inside the roundhouse on the same day. Roundhouses were not common on the Southern, the nearest being the partial roundhouse at Horsham. Certainly as loco sizes increased so the accommodation for such locos was limited. The depot also suffered badly from bombing in WW2. Even so for a time locomotives considered worthy of or scheduled for preservation were stored here. This included a member of the 'Z' class 0-8-0T class which regretfully did not end up being saved.

Opposite bottom; Looking out from the depot to the yard.

This page, top: Stored pending preservation, which in this case was successful, No 30777 *Sir Lamiel*. Recorded on 8 April 1964, the engine had been taken out of use more than two years earlier on 21 October 1961.

This page, middle: M7 No 30133 waiting its fate at Fratton on 8 August 1965. Withdrawn back in March 1964 might this too have been on someone's pending list? In the end it was not to be and No 30133 was scrapped.

This page, bottom: WR Castle 4-6-0 No 5050 *Earl of St Germans* deliberately hemmed in at Fratton on 20 July 1963. Officially barred from the SR, lines to Portsmouth, the engine had arrived on a special working but was not allowed back - A pair of Q1's 'standing guard' in front. It was eventually returned to the WR light engine but only survived in service a few more weeks being withdrawn on 20 August 1963. Excursions from the WR usually saw the Castle removed at Reading, Basingstoke or Salisbury, but examples of the class have also been noted at Eastleigh and Bournemouth. In the case of the Bournemouth arrival, the shed master is reported to have said 'well it arrived here so it can just go back!'.

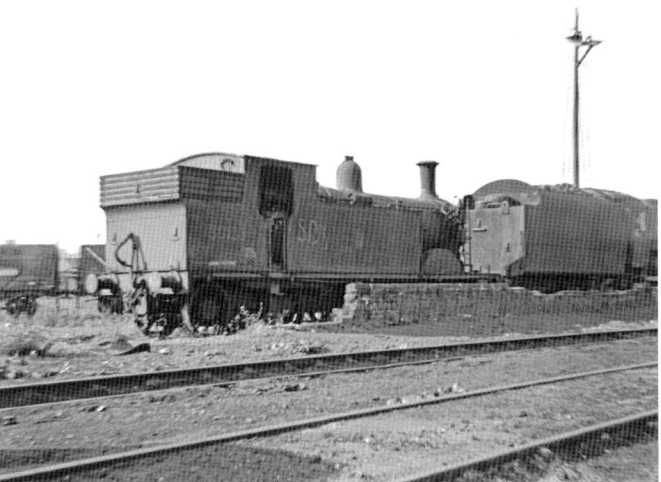

THE ELHAM VALLEY LINE—
By A. Earle Edwards.

IN PARLIAMENT
Mr. J. B. White, who called for a statement on the reason for closing the Elham Valley line of the Southern Railway, was informed by Mr. A. Barnes that the line was officially closed to passenger and goods traffic on June 16. The number of passengers using the line averaged only about a dozen a day in each direction, and the district was well covered by bus services. Parcels and small traffic could be delivered from Canterbury and Shorncliffe stations. The railway company would continue to provide facilities on the branch for dealing with the small quantity of full truck load goods traffic until satisfactory alternative arrangements had been made.

The Elham Valley Line, which was entirely closed for public use on June 14, 1947, connected the Ashford-Canterbury West Line with the London-Dover Main Line by means of junctions at Harbledown (Canterbury) and Cheriton (Shorncliffe).

It was constructed as a double line by the South Eastern Railway Company, who took over the powers to build in 1884 from the Elham Valley Railway, which was authorised in 1881 and brought into use in two stages as under:—

Cheriton Junction Box	
Lyminge Station	
Elham Station	July 4, 1887
North Elham Crossing Box	(Stage 1)
Barham Station	
Length —	
16 miles	
15 chains.	Bishopsbourne Station
	Bridge Station
	Canterbury South Station July 1, 1889
	(formerly known as South (Stage 2)
	Canterbury)
Harbledown Junction Box	June 6, 1889

The line from Cheriton Junction to Shorncliffe was single and laid in parallel with the down main line, but when Stage 2 was opened, a junction was made with the up main line on June 6, 1889 at Cheriton and the single line became the down Elham Valley Line.

Intermediate sidings were opened at Ottinge between Elham and Lyminge and Wingmore between Barham and Elham on February 6, 1888, whilst the Elham Valley Brick Company's sidings between Barham and Elham was brought into use in July 1890, but was abandoned about 40 years ago. They were all located on the down side of the line, the down direction being recognised as from Canterbury to Shorncliffe.

The line served a picturesque agricultural area and the windmill at Barham was a distinctive landmark. Broome Park (Barham) was at one time the seat of the late Lord Kitchener.

Sheep Fairs were regularly held at Lyminge and in connection therewith, a special loading dock and pens were provided at the Cheriton Junction end.

No outstanding engineering works existed, but there were a number of embankments and deep cuttings at various points and two tunnels, Each End (97 yards) between Cheriton Junction and Lyminge and Bourne Park (330 yards) between Bishopsbourne and Bridge. The bridges were numbered from Harbledown Junction to Cheriton Junction (2029-2110) and in June, 1922 the mile posts, which were positioned on the up side from Shorncliffe to Harbledown Junction were reversed and put on the down side and the mileage taken from

1. Harbledown Junction.
2. Canterbury South Station.
3. Bridge Station.
4. Bishopsbourne Station.

[*Photos. by A. Earle Edwards.*]

Kent Branch Closes after 60 Years
Harbledown Junction to Shorncliffe.

This branch line was the last on the former S.E. & C.R. to retain some of the signals with the red and green spectacle glasses fitted to the post about a foot below the semaphore arm, the green glass being of a distinctive bluish tint. Those at Barham up distant, North Elham up distant, Elham and Lyminge up and down distants survived until standardised when renewed between 1924 and 1931.

Up to about 20 years ago a number of double faced parish boundary posts were located at various points along the line; an unusual feature, viz:—

Thannington-St. Mildred.
St. Mary Bredin-Nackington.
Nackington-Patrixbourne.
Patrixbourne-Kingston.
Kingston-Barham.

At the present time many rail chairs are still in position in the sidings at Bishopsbourne and Bridge stamped S.E.R. and S.E. & C.D.R. respectively (not S.E. & C.R.). The Central Electricity Board's grid transmission lines span the railway between Each End Tunnel and Cheriton Junction.

Cheriton Halt although located between Cheriton Junction and Shorncliffe was served by Elham Valley trains and was opened on May 1, 1908. It was closed from December 1, 1915 to June 14, 1920, and again on January 31, 1941 until October 7, 1946, and has now finally closed in connection with the withdrawal of the Elham Valley Service.

On October 25, 1931, the line from Harbledown Junction to Lyminge was singled, the signal boxes between Canterbury South and Elham abolished and Tyer's key token system of signalling introduced, the portion from Lyminge to Cheriton Junction remaining double. The distant signals remained at each station to act as location indicators. Lyminge signal box was abolished on May 1, 1937, and a combined booking office and signal box brought into use on the up platform.

The line between Harbledown Junction and Lyminge was taken over by the military and the passenger train service withdrawn between Canterbury West and Lyminge on and from December 1, 1940. The local service from Dover to Lyminge was subsequently abandoned on May 1, 1943 and restored on October 7, 1946. During the military occupation certain alterations and additions were made to various sidings and the freight service was operated by Railway Operating Troops using W.D. (G.W. type) engines. The passenger train service before 1940 was provided at irregular intervals between Dover or Folkestone and Canterbury West and in some cases the trains were projected on to Ramsgate or completed the circle to Dover via Minster and Deal. "H" Class tank engines and standard Eastern Section 3 coach sets were used. Many years ago the service was supplemented by a steam rail car running between Dover and Elham, but in 1916 the service was restricted to the summer months only and finally withdrawn.

At the time of closure, the normal passenger service consisted of six trains Saturdays excepted, and seven trains Saturdays only, in each direction between Lyminge and Shorncliffe. Folkestone or Dover Priory, with one freight train daily, or as required, in each direction between Folkestone Junction and Canterbury West serving all stations and sidings.

The last official services over the line, apart from trains for engineering purposes and conveyance of odd wagons of full truck load traffic, were operated as follows:— *(continued on next page)*

5. Barham Station.
6. Elham Station.
7. Lyminge Station.
8. Cheriton Junction (Shorncliffe).

[*Photos. by A. Earle Edwards.*]

The Elham Valley line
A Earle Edwards
(at the suggestion of Peter Clark)

In Southern Times No 6 we published some notes on Arthur Earle Edwards. This prompted a response from Peter Clark and a suggestion we might care to include an article by Mr Edwards on the Elham Valley Line that appeared in the *Southern Railway* magazine for August 1947. The article is certainly worth reproducing but alas the images that accompanied it were not, hence we have included a few mostly more recent views. Worth mentioning is that there exists an excellent line history of the route (two editions) by Brian Hart and published by Wild Swan. The article by Mr Earle Edwards commences:

"In Parliament, 'Mr J B White (John Baker White, Conservative, Canterbury constituency - Ed), who called for a statement on the reason for closing the Elham Valley line of the Southern Railway, was informed by Mr A Barnes (Minister of Transport - Ed.) that the line was officially closed to passenger and goods traffic on 16 June. The number of passengers using the line averaged only about a dozen a day in each direction, and the district was well covered by bus services. Parcels and small traffic could be delivered from Canterbury and Shorncliffe stations. The railway company would continue to provide facilities on the branch for dealing with the small quantity of full truck load goods traffic until satisfactory alternative arrangements had been made.

The Elham Valley Line, which was entirely closed for public use on 14 June 1947, connected the Ashford - Canterbury West Line with the London - Dover Main Line by means of junctions at Harbledown (Canterbury) and Cheriton (Shorncliffe). It was constructed as a double line by the South Eastern Railway Company, who took over the powers to build in 1884 from the Elham Valley Railway, which was authorised in 1881 and brought into use in two stages as follows:

Stage 1: 4 July 1887.
Cheriton Junction Box
Lyminge Station
Elham Station North Elham Crossing Box Barham Station

Stage 2: 1 July 1889.
Bishopsbourne Station
Bridge Station
Canterbury South Station (formerly known as South Canterbury)
Harbledown Junction Box - 6 June 1889

Total length 16 miles 15 chains.

The line from Cheriton Junction to Shorncliffe was single and laid in parallel with the down main line, but when Stage 2 was opened, a junction was made with the up main line on 6 June 1889 at Cheriton and the single line became the down Elham Valley Line.

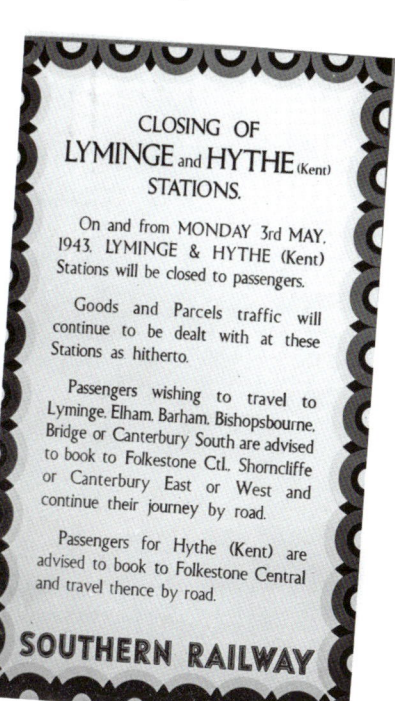

Opposite top: Between 1941 and 1944, two 12 inch mobile guns were stabled on the former Elham Valley line in Kent. Officially referred to as 'Boche-Buster' weapons, these huge machines were only able to be fired when the barrel was parallel with the track. Other Southern lines in the area Dover / Folkestone area were used in similar fashion, although as seen here, when fired at right-angles, some degree of anchorage was also required. Brian Hart refers to the first firing in his excellent 'Elham Valley' books commenting, 'Although the villagers had been warned to open all their windows, considerable damage was caused by the shock wave which brought down a number of ceilings. Subsequent firings were carried out near World's Wonder bridge and at Lickpot bridge and it was on one of these trips that an officer in charge on cautioning his men to be mindful of low bridges was promptly knocked unconscious himself as the gun passed underneath South Barham bridge.. .... Due to the tremendous force exerted during firing, the platelaying gang were regularly engaged re-aligning and maintaining the track after each exercise. One morning on a tour of inspection, platelayer Jim Weight, travelling up from Barham on the motor trolley, found the 'Boche Buster' bearing down on him. Needless to say, he broke the record for manoeuvring a reverse.' Corbis Images HU060968

Opposite bottom: The `1947 SR Magazine article (there was a further small amount of text on the following page of the magazine).

CLOSING OF
LYMINGE and HYTHE (Kent)
STATIONS.

On and from MONDAY 3rd MAY, 1943, LYMINGE & HYTHE (Kent) Stations will be closed to passengers.

Goods and Parcels traffic will continue to be dealt with at these Stations as hitherto.

Passengers wishing to travel to Lyminge, Elham, Barham, Bishopsbourne, Bridge or Canterbury South are advised to book to Folkestone Ctl. Shorncliffe or Canterbury East or West and continue their journey by road.

Passengers for Hythe (Kent) are advised to book to Folkestone Central and travel thence by road.

SOUTHERN RAILWAY

SOUTHERN TIMES

Intermediate sidings were opened at Ottinge between Elham and Lyminge and Wingmore between Barham and Elham on 6 February 1888, whilst the Elham Valley Brick Company's sidings between Barham and Elham were brought into use in July 1890, but were abandoned about 40 years ago. They were all located on the down side of the line, the down direction being recognised as from Canterbury to Shorncliffe.

The line served a picturesque agricultural area and the windmill at Barham was a distinctive landmark. Broome Park (Barham) was at one time the seat of the late Lord Kitchener. Sheep Fairs were regularly held at Lyminge and in connection therewith, a special loading dock and pens were provided at the Cheriton Junction end.

No outstanding engineering works existed, but there were a number of embankments and deep cuttings at various points and two tunnels, Ethinghill (97 yards) between Cheriton Junction and Lyminge and Bourne Park (330 yards) between Bishopsbourne and Bridge. The bridges were numbered from Harbledown Junction to Cheriton Junction (2029 - 2110) and in June, 1922 the mile posts, which were positioned on the up side from Shorncliffe to Harbledown Junction were reversed and put on the down side and the mileage taken from Harbledown Junction to Shorncliffe.

This branch line was the last on the former SECR to retain some of the signals with the red and green spectacle glasses fitted to the post about a foot below the semaphore arm, the green glass being of a distinctive bluish tint. Those at Barham up distant, North Elham up distant, Elham and Lyminge up and down distants survived until standardised when renewed between 1924 and 1931.

Up to about 20 years ago a number of double faced parish boundary posts were located at various points along the line; an unusual feature, viz:

Thannington- St. Mildred
St. Mary Bredin - Nackington
Nackington - Patrixbourne
Patrixbourne - Kingston
Kingston - Barham.

At the present time many rail chairs are still in position in the sidings at Bishopsbourne and Bridge stamped SER and SE&CDR. respectively (not SE&CR). The Central Electricity Board's grid transmission lines span the railway between Each End Tunnel and Cheriton Junction.

Cheriton Halt, although located between Cheriton

Top: Etchinghill tunnel.

Centre: Lyminge station platform side, modified since closure and with the chimneys removed.

Bottom: Platform remains at Elham.

Junction and Shorncliffe, was served by Elham Valley trains and was opened on 1 May 1908. It was closed from 1 December 1915 to 14 June 1920, and again on 31 January 1941 until 7 October 1946, and has now finally closed in connection with the withdrawal of the Elham Valley Service.

On 25 October 1931, the line from Harbledown Junction to Lyminge was singled, the signal boxes between Canterbury South and Elham abolished and Tyer's key token system of signalling introduced, the portion from Lyminge to Cheriton Junction remaining double. The distant signals remained at each station to act as location indicators. Lyminge signal box was abolished on 1 May 1937, and a combined booking office and signal box brought into use on the up platform.

The line between Harbledown Junction and Lyminge was taken over by the military and the passenger train service withdrawn between Canterbury West and Lyminge on and from 1 December 1940. The local service from Dover to Lyminge was subsequently abandoned on 1 May 1943 and restored on 7 October 1946. During the military occupation certain alterations and additions were made to various sidings and the freight service was operated by Railway Operating Troops using W D (GW type) engines. The passenger train service before 1940 was provided at irregular intervals between Dover or Folkestone and Canterbury West and in some cases the trains were projected on to Ramsgate or completed the circle to Dover via Minster and Deal. 'H' Class tank engines and standard Eastern Section 3-coach sets were used. Many years ago the service was supplemented by a steam rail car running between Dover and Elham, but in 1916 the service was restricted to the summer months only and finally withdrawn.

At the time of closure, the normal passenger service

Top left: The restored Barham signal box. All that is needed now is a restored 'boche-buster'.... .

Top right: Bishopsbourne.

Centre: Bishopsbourne.

Bottom: Bourne Park Tunnel.

SOUTHERN TIMES

consisted of six trains Saturdays excepted, and seven trains Saturdays only, in each direction between Lyminge and Shorncliffe, Folkestone or Dover Priory, with one freight train daily, or as required, in each direction between Folkestone Junction and Canterbury West serving all stations and sidings. The last official services over the line, apart from trains for engineering purposes and conveyance of odd wagons of full truck load traffic, were operated as follows:

Freight. 13 June 1947. 1.18pm Canterbury West to Folkestone Junction. '01' Class Engine No 1381. Driver Bailey, Fireman Anderson, Guard Howell. All Folkestone Junction men.

Passenger, 14 June 1947. 9.0pm Lyminge to Folkestone Junction. 'H' Class Engine No 1531. Driver R Wells, Fireman G. Twyman attached to Dover Marine. Guard Neville attached to Folkestone Junction.

Canterbury South and Bridge Stations were under the control of the Stationmaster at Canterbury West and the remainder in charge of a resident Stationmaster at Elham, Mr Caple, who for a number of years was well-known in the West of England and Gomshall areas."

Canterbury South, 70 years since its last train.

Treasures from the Bluebell Railway Museum
Tony Hillman

The Southern Railway Horse Department

The Southern Railway carried out a census of staff on 5th March 1931. The document, a copy of which is in the Bluebell Archive, was produced by the General Manager's Office on 1 September 1931 and runs to 639 pages.

The Horse Department had a total of 216 adult male staff, one junior male and one female employed in 25 Depots.

Carters, who drove the carts, were employed in the Traffic Department. There were 679 spread across the network. About 200 of these were stationed at locations not included in the Horse Department. For example, Lewes is not recorded in the Horse Department as having a stable but has three carters. Lewes had numerous stables so, presumably, the Southern Railway had an agreement with one of them.

Bricklayers Arms was the largest recorded depot having 72 staff. The two Veterinary Surgeons employed by the Southern Railway were based here. Another position that was only recorded here was that of Chaff Cutter. These fifteen staff cut straw and hay into short lengths for mixing with horse feed. The one carter employed by the Horse Department was here and, presumably, transported horse feed around the London Depots.

Next largest Depot was Nine Elms, 61 staff, followed by Blackfriars, 27 staff, and Willow Walk (Brighton), 21 staff.

Other Stations/Depots with stables were: Aldershot, Ashford (Kent), Bournemouth Central, Brighton, Canterbury West, Chatham, Clapham Junction, Dartford, Dover Priory, East Croydon, Eastbourne, Folkestone Junction, Gillingham (Kent), Hastings, Maidstone West, Margate, Tonbridge, Tunbridge Wells Central, Southampton Terminus and Woolwich Arsenal

Southern Railway horse and cart at Lewes on Friday February 23 1951. The tarpaullin is marked 'B.R.(S)'.
All images courtesy Bluebell Railway Museum

SOUTHERN TIMES

	Vet. Surgeon	Chief Clerk	Clerk	Foreman	Inspector	Carter	Chaff Cutter	Stableman-in-Charge	Stableman	Working Foreman	Farrier	Harness Maker	Clerk - Female	Farrier's Boy	Chauffeur	Total
Bricklayers Arms	2	1	2	3	2	1	15	2	32	1	3	5	1	1	1	72
Nine Elms				4					48		6	3				61
Blackfriars				2				2	19	1	2	1				27
Willow Walk				2				1	15		2	1				21
Other stations/depots				3				6	28							37
Total	2	1	2	14	2	1	15	11	142	2	13	10	1	1	1	218

Cartage Artifacts on display in the Museum.

Left: Again at Lewes on the date mentioned on the previous page. Unfortunately the name of the man and horse are not reported.

Bottom; One of the shelves in the display case shows a model horse and cart which measures about 18" long. It is assumed that the model was an advertising display in a Goods Depot or similar. A set of horse brasses and a "Cleansed & Disinfected" wagon label. One can only imagine what sort of job cleaning a horse box was.

Opposite, top left: The original location of this plate is unknown. It could have been located at the entrance to a Goods Depot, such as Lewes or on a cab road at a station e.g. Brighton.

Opposite, middle left: We believe that this brass tablet was used as a permit to allow Hanson Cabs to use Brighton Station cab road. It is about 6" wide with a loop for use of a leather strap.

Opposite right: A 1930s Time Recorder Clock as used by cartage and lorry railway workers, a kind of early tachograph. It would have been attached to a cart to record movement.

Contains a wind-up clock which records movement on a circular paper disc. The time that movement started and ceased would be marked. The disc showed 7 days movement. The recorder is stamped SR 47 and is also marked as last being serviced in February 1961.

Re-evacuation of school children from Kent and Sussex. An Update.

In Southern Times 6 the trains re-evacuating children away from the South Coast were described. At the time of writing no document could be found to complete the picture. To where in Wales did the trains go? In one of the boxes in the Archive just starting to be catalogued, the document to complete the story turned up.

The seven trains via Reading went to Swindon. Two went to Brecon (LMS) via Gloucester, two to Fishguard and one to Garnant, Carmarthen and Narberth. Stops were made to drop of children at various stations en route, e.g. Hay, Talgarth and Llanelly.

The seven trains via Salisbury continued via Westbury, Bath and Newport. Destinations included Carmarthen, Llandovery, Haverfordwest, Llandebie and Newcastle Emlyn.

Don't forget, the Bluebell Museum at Sheffield Park has regular special exhibitions devoted to particular themes. Recently these have included
Railway Women
and also
Trainspotting.

The 1948 Southern Region Locomotive Building Programme
The nightmare...

In issue 8 of Southern Times we dealt with the perceived steam locomotive requirements for motive power on the Southern Region in the immediate post-nationalisation era.

With hindsight, it is all too easy to say the railway should perhaps have done 'x' and 'y' but there were a number of factors to take into consideration.

Principal amongst these was the need for a large tank engine capable of working passenger services and as recounted in the last issue, this need was filled initially by the construction of a number of LMS design 2-6-4T engines at Brighton and later numerous BR 80xxx series tank engines at the same works.

Slightly beyond the period in question, a further urgent need would occur when, with the end of the oil-burning programme a number of Drummond 4-4-0 tender engines were not reinstated to coal burning. Instead they were withdrawn en masse, leading to a shortage of motive power on secondary passenger services, particularly those previously worked from the depots at Basingstoke, Eastleigh and Fratton. This resulted in the temporary transfer of 4-4-0 types from the Eastern section with the shortfall only really resolved following the building of the BR Standard type tender engines of the 76xxx design and which it may be said, were welcomed with open arms compared with the aged remaining Drummond types. (Sorry Southern aficionados but we have to be truthful.)

Having then set the scene as to the motive power situation as it developed in the years immediately after 1948, we now come back to the various minutes of the committee discussing the locomotive building programme and with specific regard to what would be No 36001 and its potential siblings.

We start on 13 January 1949. The first 'Leader' is already 18 months late and the Public Relations Department at Waterloo is starting to get enquiries over the construction of the engine. Cuthbert Grasemannn, in charge of the PRO at Waterloo has referred the enquiry to BRB - via Elliot (Chief Regional Officer), the following response received at Waterloo from A J Pearson at Marylebone.

"Grasemann will no doubt have told you of his conversation with Barrie (?) yesterday in regard to the publicity when the first of these locomotives is completed. The position is that we have already had a request from British Movietone News to take shots of the locomotives under construction at Brighton and from the BBC at Bristol to make recordings of other work, which they say is being done at Eastleigh.

Whilst we have no desire to hamper legitimate Press publicity for any new development, it appears from the enquiries already made that there is some public impression that these engines are, in fact, the first to be constructed as a prototype for British Railways as a whole. This is not, of course, the case and you may agree that it will be desirable in any publicity to make the position clear and to indicate that the somewhat revolutionary design is one decided upon by the former Southern Railway.

I would be grateful to have your views as to how the publicity should be controlled so as to avoid giving any false impression and at the same time to secure adequate results. My own feeling is that the case could probably be met by avoiding advance publicity prior to the first engine of the type coming out of the works, when a general press inspection could be held; having regard to the policy factors involved you will no doubt be good enough to let us see in advance the draft of whatever you propose to issue to the Press, so that the technical aspects may be examined by the responsible Member of the Executive."

Again with hindsight this raises a number of questions and similarly warrants a few comments but before getting to that stage let us report on the reply from Elliot to Pearson dated the following day.

"Thanks for your letter of yesterday". Riddles spoke to me at the Railway Executive yesterday and told me he had spoken to the Chairman about this matter. I told him that I am in agreement with the Chairman and Riddles that the best thing to do is for Grasemann to stave off any publicity, if he can do so, until the first engine is completed and has run its trials. Then, on the day before it is put into traffic, it will be best to let the Press see it and give them a story which could be agreed previously of course with Mr Riddles and yourself, on

It is unlikely anyone reading this will not know what No 36001 looked like, whilst finding an unpublished view of the engine is getting increasingly difficult. Suffice then for this undated image of No 36001 returning to Brighton, sometime in the autumn of 1949 the oscillating gear to the sleeves has already been removed and see hauling a test train. At some point 'Sillimentia' kiln firebricks were fitted, these found to have a service life of just 1,468 miles. *Neville Stead NS200949A / Transport Treasury*

the lines that it is an experimental locomotive ordered under the auspices of the Southern Railway Board and does not commit British Railways to continuance of the type, although its performance will be matched closely and, if it justifies itself, full consideration will be given to its merits.

I am sure that Bulleid would be in agreement with this and I will have a word with him and I am sending a copy of this letter to Grasemann."

It is hard to comment objectively on the response from Pearson, absolute incredulity springs to mind. On the one hand BR are leaving it open to take the credit - if it turns out to be a success - and yet also paving the way to blame history, the Southern Railway, should it turn out to be a failure. 'Spin' again comes to mind - 'twas ever thus..... .

One point to note also is that somehow the story of No 36001 and to be honest of course it was indeed revolutionary for the time, had leaked out to the press, Movietone, BBC and perhaps others. (Deliberate or accidental...?) Not surprisingly they then sensed a story but were effectively kept in check with the promise of 'jam tomorrow'; perhaps when tomorrow did come there were more interesting stories to report upon. Even so the only known movie footage of the engine is a short sequence in its unfinished state concentrating on the bogies in steam possibly taken at this time. Meanwhile perhaps the BBC decided their own proposed visit to Eastleigh was just not worthwhile.

Two weeks later on 26 January Bulleid himself writing to Elliot from Brighton adds his own comments, 'I am obliged to you for your letter of the 14 January enclosing copy of your letter of the same date to Mr Pearson and copy of his letter of the 13 January to you.

"I would, however, suggest that the story cannot be agreed with Mr Riddles but should be left to Mr Grasemannn. Mr Riddles is in no way involved in the technical design of these locomotives and to bring him in would give the impression that the Railway Executive was in some way concerned in the design of such locomotives.

In your letter to Mr. Pearson you suggest that there be no publicity until the first of the engines has run its trials. This, I suggest, is an impossibility under present day conditions. As a great number of people know these engines are under construction, as soon as the first is steamed and leaves the Works for its first trial, this fact will be given much prominence."

Grasemann agreed with Bulleid, responding to Elliot on the same day. *"I entirely support Mr Bulleid's views that once the engine starts its trials it is impossible for publicity to be withheld, as in these days when railway enthusiasm ranges so high amongst the public, photographers and 'spotters' will be lining the various stations where they anticipate this engine will appear.*

'If we are to obtain the right publicity, particularly photographic, we must have freedom of action immediately the engine leaves the works."

Now was this the 'professional' Grasemann or the man who perhaps watching the success of his former junior employee Ian Allan, had come to realise the enthusiastic following that existed for railways? (Whilst initially dismissive of the enterprising Ian Allan in producing what was his first 'ABC' book in 1942, Grasemann ended up writing a booklet 'Round the Southern Fleet' for Ian Allan in 1946!)

SOUTHERN TIMES

Meanwhile some 175 miles from Brighton where Leader was under construction, the *Pontypool Free Press* on 20 May 1949 published a short report that would have ramifications on the Southern Region.

In itself it was an innocent enough report, simply that a local resident had enjoyed a ride in the cab of the Brighton Belle and then been taken on a tour of Brighton Works.

"Mr John Drayton, of 13, Sycamore Road, Griffithstown, recently rode in the cab of the 9.00am electric Pullman car express from Victoria (London) to Brighton, where he was conducted on a tour of Brighton Locomotive Works to view a new and revolutionary type of steam engine under construction, now nearing completion.

'Conventional side rods are eliminated. The driving wheels are coupled by means of chains, and the running gear is enclosed and lubricated in oil baths. There are two six-wheel bogies, and six cylinders (three for each unit). This very striking machine can be driven from either end, giving a completely unobstructed view of the track ahead.

'The performance of this unusual machine of the 'Leader' class will be watched with the keenest interest by locomotive men the world over. Mr Drayton rode back from Brighton in the motorman's cab of the famous 'Brighton Belle' all-Pullman electric express, which covers the 51 miles to London in 60 minutes."

So, on the face of it an innocent report in a local newspaper, Mr Drayton seemingly having also gleaned some basic technical information to enhance his experience but BR at Marylebone were clearly very touchy about the whole build and a week later on 27 May Pearson at the Railway Executive was in contact with Elliot at Waterloo over the newspaper report. (How Marylebone had heard of the report is not certain - more leaks perhaps - no pun intended.)

".....I attach copy of an extract from the 'Pontypool Free Press' of 20 May 1949, from which it appears that some journalist or member of the public has been given an opportunity to inspect this class of locomotive and to obtain information about it whilst under construction at Brighton Works. While the newspaper itself is not an important one, the principle that such facilities should have been obtained would seem in the circumstances to be somewhat unfortunate, and I shall be glad to have your observations."

Elliot made his own enquiries reporting back, *"Phoned Brighton CME on 30 May to know details of Mr Drayton's visit to Brighton. Replied 31 May.*

Mr Drayton is an engine driver on the Western Region at Swindon and sent Mr Bulleid a very good pen sketch of a Merchant Navy locomotive. He also requested permission to ride in a motorman's cab and visit Brighton Works.

As Mr Drayton was obviously interested in Railways Mr Bulleid arranged for the issue of a cab pass and the visit to Brighton. Mr Drayton subsequently wrote to Mr Bulleid enclosing a report on the visit with a brief reference to the Leader class engines and asked permission to issue it in the 'Locomotive Express. After satisfying himself that the article contained nothing which had not previously been published, Mr Bulleid agreed to the publication."

Note - we have been unable to locate any contemporary publication under the name 'Locomotive Express'. An alternative is that this was a (semi) regular / occasional section of a newspaper. The name Brian Drayton also comes up in two books penned - we assume - by the same man on footplate experiences. 'On the Footplate - Memories of a GWR Engineman' Pub. D Bradford Barton, 1976, and 'Across the Footplate Years', pub. Ian Allan 1986.

Returning to 1949 and the present writer cannot help thinking that this was deliberate back-door publicity for the engine engineered by Bulleid. Drayton's request for a cab ride suitably manipulated for his own purpose.

Leader, as we know, finally emerged from Brighton in June 1949, Bulleid contacting Elliot at Waterloo on 31 May with a draft press release previously sent to the Railway Executive a few days before. Entitled 'The Double Bogie Steam Engine' it reads, *"At the time of the grouping of the Southern Railway constituent Companies, there were a large number of tank engines in service on each of the three Sections. These tank engine services had been built up due to the suburban type of traffic. The reason for building tank engines is to overcome the necessity of turning the engine round at the termination of a run, ready for the return journey, more particularly with an engine used for suburban traffic. In addition to this, on the London, Brighton and South Coast Railway, the Chief Mechanical Engineer responsible for the construction of new engines considered that it was desirable to build tank engines so as to obviate the necessity of hauling the tender for the purpose of carrying water and coal.*

Due to the continuous growth of electrification on the Southern, the use of the tank engine became more limited as the suburban traffic was largely undertaken by multiple unit stock over the electrified lines. This circumstance led to there being more tank engines available than there were duties for them to perform and, indeed, one class of engine was converted from a tank engine to a tender engine as sufficient duties could not be found for the economic use of the type. (In fact there were two, the 'River' class converted into the 'U' class and the Brighton 'L' type which became the 'N15X - Ed.)

It is now considered by the Traffic Department essential to have new passenger tank engines available for certain services, particularly where a quick turn-round is necessary and also for such duties as Exeter to Exmouth, Bournemouth Central to Swanage, Brookwood, (or similar outlying stabling grounds), to Waterloo. To meet these conditions, the Chief Mechanical Engineer of the Southern Region decided to build a tank engine suitable for all ranges of traffic from the express passenger trains down to the more heavily loaded of the suburban type of train still requiring to be hauled by steam locomotives. It was decided that the fundamentals which should be observed in the designing of the tank engines were...."

Here sadly the second (and perhaps subsequent pages) are missing from the official file but we may have no doubt they would have contained the basic criteria and purported advantages of Leader which have been outlined many times before.

Previously in this piece we referred to the anticipated publicity the railway might offer and expect following the introduction of the first engine into service. 'Into service' never of course occurred and instead it appears it was left to just one newspaper in an undated issue to report as under. 'Star' Reporter. On the result of inspections now being carried out at Brighton by British Railways depends the introduction of what railway officials describe as a 'revolutionary' type of locomotive.
"A few people on the Brighton - Three Bridges line saw the train during a recent two-day running test. They may have thought it the usual electric train running without carriages because there are no funnels to be seen and the driver sits in a cabin with an unrestricted view as in the electric train. The 'Leader' engine can be driven from either end, and the fireman works in a compartment in the middle of the locomotive which is connected with the two driving cabins by a corridor.

The driver works his train by remote controls. The object of the Leader is to facilitate a quick turn at the end of a journey. 'We are now seeing how it stood up to the trial runs,' said a SR official. 'We want to see how much fuel it used, find out what the driver and firemen, thought of it, and examine it mechanically to estimate wear and tear. 'This is an idea on which we have been working for years. If it comes into use as a result of these tests it may be seen first of all on the London-Brighton line."

An unknown official had written in pencil, *"It certainly won't"* against the words 'London - Brighton line'!

Matters then go quiet for the next three months when Bulleid writes to Elliot at Waterloo using the heading 'Leader class locomotives, Press Inspection', *"In reply to your letter of the 15 August, I feel that the particular trouble which has been experienced has been overcome, and the provisional date for the run with the engine painted in its proper colours, ready for press inspection, would be about the middle of September.*

It will be necessary for me to confirm that at a later date, which I will do as soon as possible."

There followed a table of costs, compiled by the Chief Costing Office within the Accountant's Department at Brighton but for whom is not stated.

As we know the Leader trials were not successful and in November 1949 Riddles postponed work on the other engines of the class, Nos 36002-5; it would never

	Total expenditure up to and including 9-10-1949		Cost of loco No 36001		Cost of locos. Nos 36002-5		Expenditure on loco 36001 since leaving Erecting Shop	
Material	76,348	2s 5d	14,144	5s 11d	62,118	14s 0d	85	2s 6d
Wages	51,240	13s 2d	14,295	14s 1d	35,616	11s 10d	1,328	7s 3d
Workshop Expences	39,673	0s 4d	10,913	7s 4d	27,980	18s 4d	778	4s 8d
Supintendence	5,124	6s 4d	1,429	11s 5d	3,561	18s 2d	132	16s 9d
Total £	172,386	2s 3d	40,783	8s 9d	129,278	2s 4d	2,324	11s 2d

Behind the scenes the new man in charge at Brighton and who had replaced Bulleid was S B Warder, by profession an electrical engineer but who nevertheless also took over responsibility for the region's steam stock.

Hardly surprising was there were issues with the springing of No 36001, caused of course by the placing of the boiler other than on the centre line of the locomotive. It was evidently felt altering the springs might be of assistance, the result of this endeavour reported to Riddles by Warder on 26 January 1950.

"Referring to your letter of the 29 December: I have to inform you that the further work requested by Mr. Bulleid on his last visit to Brighton has now been completed.

You are aware that this included changing the springs, which when carried out was unsatisfactory since the new springs were too weak to carry the load. Subsequently, a matched set of the original springs was refitted, to which Mr Bulleid gave his approval, as it was the only immediate alternative.

A preliminary weighing with the new fire bricks has been carried out at Brighton, which shows that the total weight with -

Coal   4 tons
Tank water   4,000 gallons
Water in boiler   1" in glass - cold.

gives a total weight average over a number of weighings of 133 tons 13 cwts.

Reported: weighings on the Brighton showed that owing to the rigid connection between the main frames and the bogies, and the resistance to the side movement of the axlebox guides, inequality of weight between the left and right sides cannot be determined with any accuracy.

The above weights are in fact, disputed by the Drawing Office here, which opinion Mr Bulleid is inclined to uphold, and in my opinion, therefore, it is most desirable that this issue be clarified and the question of the accuracy of the weight put beyond doubt.

Mr Bulleid is of the opinion that correct weights cannot be obtained either at Brighton or Eastleigh, and proposes that the engine should be run in the following condition:-
'Springs, brickwork, tanks, etc., as at present fitted, but

Coal not to exceed 3 tons.
Tank water to be limited to 3,000 gallons.

Our estimated weight in these circumstances is approximately 127 tons. (At Eastleigh on 16 December 1948, the weigh table there showed a difference of 9 tons 6 cwt. between the two sides; the heavier side naturally being that towards which the boiler was offset.)

As a set of the original stiff springs has been fitted, it is particularly desirable that a preliminary run should be made to settle the springs as far as possible before an attempt is made to equalise axle weights for the final weighing. I shall be glad to have your further instructions in this matter, having regard to the fact that you require two or three days notice before the locomotive can make the journey to Eastleigh, and I presume that you will make the necessary arrangements for Mr Train to suitably instruct Mr Robertson. (No known relation - Ed)

In conclusion, I would mention that every visible assistance has been given to Mr Bulleid, and the work that has so far been carried out has been done to his instructions."

Matters now move forward beyond January 1950, past the time of its transfer to Eastleigh, the crank axle failure, subsequent repairs and then the dynamometer trials. In fact to the time when Leader lay cold in a siding at Eastleigh awaiting its fate.

The sender is unknown but we do know the recipient, again Mr Riddles. Written on 18 October it refers to correspondence not in the file but continues, *"Referring to my letter of the 2 October and conversation on 4 October at Brighton; it is your intention that a further trial should be made with this locomotive hauling a train loaded up to the capacity which Mr Bulleid had in mind that it would regularly be capable of working. It was ascertained that this load was 420 tons, but as there was such a difference between this and its previous maximum of 325 tons, which it had worked reasonably successfully, it was not thought advisable to load the locomotive with this amount and therefore, a trial run took place last night between Eastleigh and Woking with a load of 430 tons.*

The train was timed reasonably in accordance with schedule and the net time between Eastleigh and Basingstoke was less than the schedule when allowance was made for a permanent way slack of 15 mph between Winchester Junction and Wallers Ash. The boiler pressure was fairly well maintained, the minimum on the bank being 195 lbs. per square inch;

but the level of water was appreciably lowered. The total quantity of coal consumed on the run was 52 cwts. and the water 3,664 gallons.

As the dynamometer car was not used on this run, I have had to assess the drawbar horsepower hours from standard resistance curves for passenger rolling stock and on this basis the coal consumption was 6.5 lbs. per drawbar horsepower hour and the water 41 lbs. per drawbar horsepower hour. As a matter of interest, one ton of coal was consumed between Eastleigh and Basingstoke, representing a firing rate of 132 lbs. per sq.foot of grate per hour.

In view of the above, is it your wish that still further trials should take place?

As you are aware, I have in hand, in conjunction with Mr Harrison, the preparation of a report on the trials carried out on this locomotive, together with certain other supplementary information which I consider you will wish to have placed on record.

I am, of course, in the hands of the dynamometer people at Darlington, but I hope to be able to let you have this report in the course of a week or so."

A week later a shortened version of the events with Leader hauling 430 tons was sent from Brighton to Elliot at Waterloo. The signature, although present, cannot be deciphered.

"25 October 1950. With reference to your letter of the 17 October, following the completion of the Dynamometer Car trials of the above engine, Mr Riddles asked me to carry out further tests with a heavier load than we had previously attempted.

Accordingly, on the night of 17 October a test was made between Eastleigh and Woking with 430 tons. Following this I wrote to Mr Riddles informing him that the locomotive had dealt with this load reasonably well and asked him whether he wished any further trial. For your information, I enclose copy of my letter dated the 18 October to Mr Riddles and I have now received a reply from him, stating that he wishes a still further test to be made with a load of 480 tons, which it is proposed to carry out on Tuesday, the 31 October.

Since the rings have been changed on this locomotive appreciably improved results have been obtained but the coal and water consumption are still approximately double those that would be expected from a modern design of locomotive. When the remaining tests have been completed I hope to be in a position to report fully on the locomotive, from the point of view of economy as well as performance."

As we know the trial with 480 tons did take place, not on the date quoted but for whatever reason on 2 November. The engine completed the run but effort needed was such as to warp the smokebox door upon examination at Basingstoke; it had simply not been secured properly before leaving Eastleigh and consequently was drawing air. Some might ask surely this would have been noticed but as ever Leader was different for a considerable amount of insulating material had been placed around the smokebox in an attempt to reduce the heat in the driver's compartment at this end of the engine. We have never found out if this was entirely satisfactory. Whatever, with all this insulation present it was not easy to see if the smokebox door was in fact securely fastened. The cure was easy, take a door off one of the incomplete engines but it was never done for as we know, once the engine returned to Eastleigh it never ran under its power again.

This is confirmed in Minute 27 of an undated meeting here under item 27 'Locomotive Building and Condemnation Programmes, 1948, 1949, and 1950', item '27a' reports, 'Cancellation of authority for the construction of five mixed traffic locomotives at a total estimated cost of £100,000', and 'Scrapping of locomotives already completed and, also, of work done on the other four'. This was approved by the BTC on 6 March 1951.

And there we might expect the whole Leader saga to end, excepting of course it did not with the accusations, we can hardly call them revelations, reported in the 'Sunday Despatch' in 1953. Except even then it did not totally lie down. Witness then a letter from H W Gunter Kaestner in Germany received at the BTC in 1958 part of which was sent on to Waterloo to be dealt with there. It reads:

"May I ask you courteously if it would be possible for you to give me also detailed information about your rare chain-driven Leader class engines? Please send me data-sheet and photo of it and tell me please what happened with these locomotives and how many have been built. This request in addition to my letter of 21 April. Hoping to hear from you I thank you very much in advance for your kindness."

The BTC added the comment to Waterloo, *"No doubt you will wish to reply direct to Mr Kaestner regarding this subject."*

Waterloo did reply, contacting Mr Sykes, in his capacity

as Chief Mechanical and Electrical Engineer of the Southern Region for his response.

We only have the letter from Sykes and not what was sent back to Germany but suffice to say it summarises well a difficult subject which no doubt was best felt buried - as indeed it was until 1985 when your present writer stumbled upon the recently released test reports at the NRM at that time - and the rest I am sure readers will know.

"In reply to your letter of the 23 May, I suggest that Ing. B W Gunter Kaestner should be advised as follows:-

There are no 'Leader' class locomotives in operation on the Southern Region of British Railways. A design was prepared for these locomotives in 1946, and an experimental prototype was built in 1949. The locomotive did not, however, go into regular service and it was subsequently broken up, as justification for a locomotive of this type ceased with the British Transport Commission's decision to proceed with the Modernization Plan, based on electrification and dieselisation.

I enclose a diagram showing the general details of the prototype locomotive, together with a photograph. The principle envisaged in the design of this locomotive was that the motion should be incorporated in the power bogies, in order that, for purposes of maintenance and repair, a bogie could be removed from the locomotive and replaced with a spare unit, in order to keep the locomotive in service. The design was also such that the locomotive could be driven from either end, in order to obviate turning at terminal stations."

Sykes concluded to Waterloo as follows, *"I would mention, however, that the policy laid down by Mr Riddles and Mr Swift was that any information requested by outside bodies regarding these locomotives should be refused, but, owing to the passage of time since the scheme to build these locomotives was abandoned, you may consider that there is now no reason why limited information as above should not be released."*

BR had their fingers publicly burned in 1953 and clearly for some years it was a difficult subject.

So endeth the saga of Leader, taken from papers at the National Archives, RAIL 1168-207. These were not available to the author at the time of his original research and whilst the basic facts of the Leader story remain as previously stated elsewhere the preceding does make for interesting additional information.
In reality, could the engine have worked? Well that is the question both engineers and enthusiasts have vexed opinions upon. The views of the present writer have been published before and I do not change them now, 'Leader did not fail us - we failed it - for failing to recognise development of the steam locomotive was still possible, but clearly not in the way Leader took us'.

A balanced view also comes from retired Eastleigh and Southern Region trained engineer John Wenyon who comments:

My own take on Bullied's sensible course of action is as follows;

1. Keep the traffic department quiet by building a reversible version of the Q1's. The unstable tender issue is a red herring in that a new tender version would have been needed anyway. Basically with a cab front to protect the crew from 70mph reverse running together with inset coal bunker sides for good vision. Then, if necessary, replace the rear wheelset with either a cartazzi axle or better still, use one of Percy Belloms' link guided pony trucks as used on the big diesel four axle bogies (Nos 10101 etc, class 40/45). Perhaps fit side buffers for guidance between the loco & Tender (use the EMU centre buffers with new springs). This would deal with the Waterloo ECS workings to Clapham & Basingstoke.

2. 'Q1R' (my class designation??)

3. The tender weight needs to be increased to give the class good braking power. There is scope to bring the axle loads up to the 18.5 tons on the loco if needed.

4. My observations watching Q1's shunting is that a finer control regulator would be beneficial (fit a smaller one as was done later on the 9F's, the local throttling of the good steam circuit did not affect performance).

5. It would be nice if power operated Ajax fire doors and electric lighting was fitted - i.e. make them a bit more civilised to live with.

6. Then build his all adhesion bogie steam loco, but keep it simple, fit a conventional boiler on the loco centre line and forget sleeve valve cylinders, I am not convinced there is any real efficiency gain available there. Build removable 3-cylinder steam engines as in the CIE turf burner 2-cylinder engines. This would enable use of smaller wheels (4' 1" maybe - as required to conform to the required wheel diameter / axle load regualtions)) with the drive gear ratio selected for the design speed of the loco, 75mph would do but 90mph is available for an express version if desired. I consider that a central

cab would be better as simpler to build and to keep the crew together. I note your point about push-pull working separating the crew, but push pull fitted locos were generally small with a good vision for the lone fireman with access to a full set of controls.(It was not unknown for the Fireman to drive the loco relying on his own vision and the drivers bell signals). If however end cabs were required then the loco would be longer and my thought about the smokebox end cab would be to leave an open space between the smokebox and cab back (insulated) for air cooling, i.e. open at the top and fit air louvres at the sides. I believe that provision for open access doors between the smokebox and front of the loco should be made to allow boiler tube cleaning etc. a bit complicated but possible. Fitting Dead-mans controls would be possible using a pedal to hold the brakes off (as developed for 20001) together with an electric bell alarm to the firing cab (which should also have regulator control and reverser control as well if possible).

7. Further thoughts I forgot to put into my letter re Leader. The demountable engine's crankshaft would be free from the stresses of being also a driving axle and gear driven gives some freedom of gear ratios thus acceleration and top speed, depending on the main duties of the loco. Assuming the continued use of chain drive between the coupled axles there might be clearance to mount the chains inboard of the wheels which might reduce stress on the driving axle, which not being a crank axle can also be generously built.

8. If the Leader was successful as a traffic machine in this or similar form, we still have to face the fact that it is a coal burning steam loco with the same thermal efficiency and coal consumption as the rest of the breed and still requiring the same manpower needs. The clear grate space and shorter boiler will make it an efficient steam producer. However, it should have a much better acceleration than conventional machines, due to its all adhesion design and high speed balanced engines.

9. Disadvantages: Potentially higher weight than equivalent power steam locos, greater length and higher first cost.

10. Bullied and Raworth. Two very similar and very clever engineers who held diametrically opposite views on railway development, who irrespective of personality differences were never going to agree on much. Bulleid came with the view that the Southern had overdone electrification in electrifying its main lines and Raworth, probably from his earlier experience working with his father on electrifying rural tramways, wanted cheap total system wide electrification .

11. Raworth was a far better project manager than Bullied, but did not have Bullied's innate skill in persuading management boards to fund his development projects. It is interesting that when Raworth applied to join the Institute of Civil Engineers it was the CCE George Elson who proposed him for membership!

12. Raworth's version of a 'Leader' type project was his 3,000V 4-rail electrification scheme for the SE.& CR with two live rails each at 1500V one + and the centre rail -. Luckily for track staff Herbert Walker squashed that plan in the pursuit of standardisation. In spite of expert electrical engineers approving the ideas even though Raworth would have reduced the voltage to 1500V (750V + and 750V -) which might have allowed some inter-running of existing southern electric trains, and, if the early Channel Tunnel scheme had been built 4-rail trains, loco led could have started in London, gone to France and on to Paris using the 1,500V overhead system there.

John concludes, 'I would like to have known Raworth'.

Do you have your own thoughts on No 36001?
Do you agree with the Southern Railway / British Railways allowing the build to proceed?
Were BR correct in subsequently cancelling the project?
What would you have done?
Tell us your opinion, we would like to hear your views.

From the Footplate

We start this issue's selection with a note from **Tony Wardle** concerning an image, well part of it actually, that appeared on **page 57 of Issue 8**. *"I was very interested to see this 1955 photo at Blackheath included what looks like one of the weed killing trains I saw when travelling from this station to school in 1961/2. I noted the formation as (all with the prefix 'DS'): No 455 brake van with date 23/ 2/ 57, Nos 1477/ 76/ 75 ex tenders marked 'water' with 'Southern on No 1477, Nos 1474/ 73 also ex tenders marked 'Conc' with a BR lion and wheel on No 1473, No 1478 another ex tender, this one marked 'Mixture' and finally Nos 470/ 471 both former utility vans for staff. Does anyone have further details of these vehicles please and what became of them?"*

On Tony's behalf we contact Mike King who has kindly responded, 'Not so straightforward as you might think! It is one of the weedkilling trains, but exactly which tenders could vary as there were several incarnations of these. The first appeared in 1931 and included four LSWR Adams tenders between two LBSCR brake vans. It is possible that a few of the tenders are seen here. These were numbered Nos 572s-575s in the departmental list - later DS572 etc.

"In 1936 a further train was converted, using ex-SECR 6-wheeled passenger brake vans and six ex-SER tenders - the tenders being numbers 1038s-1042s (plus one more I haven't found in the departmental list. Again, some of the tenders might be visible here as there was a degree of mix and match to these. They generally ran as pairs, but the pairs could be swapped around.

After World War 2 a further pair of trains were converted, both using two SR utility vans and one SR 15-ton brake van - and three of the vehicles are clearly seen here. These were almost new SR vans Nos 1787, 1791, 2127 and 2164, which became (respectively 466s, 469s, 470s and 471s (later DS466 etc) and 15-ton brake vans Nos 455s and 456s (ex- 55719 and 55711 respectively), all being converted in 1947/48. Here we are probably looking at DS455 and DS470/471 as these generally ran together. The other three ran with tankers that had new cylindrical bodies placed on old tender frames and these were numbered 458s-460s and 462s-464s. Six more tender conversions were also done, these being 1473s-1478s (later DS1473 etc.). Some were LSWR, others SECR and again a pair or two might be here.

However, there was a further change in 1957, when replacement Nos DS1473/74/75 were done and these replaced the original DS1473-75. Obviously, these are not in the picture, but may be some of those your correspondent recorded. By 1967 at least one train was running with a mix of both tenders and tankers, so a further set of swaps had taken place. Withdrawal of one set took place in 1968, but the other (not necessarily all from the same set) carried on into the 1970s. I think brake van 455s/ DS455 is now preserved on the East Kent Railway at Shepherdswell."

Further thoughts from readers are welcome.

Next from **Roger Whitehouse** on **ST8** reference **EMUs, headcodes and the duties of the N15X class engines.**

"The latest issue arrived this morning, full of fascinating stuff as usual. Although I haven't yet read it in detail I do have some immediate comments on the captions to EMU photographs.

Page 42: the headcode appears to be 03 (with a zero) not G3. As far as I know, the only stock able to mix letters and digits in Southern style headcodes were locomotives fitted with roller blinds. 4747 appears to have stencils. The WTT dated 1/5/72 instructs "Empty trains from Clapham Junction, Durnsford Road, and Wimbledon Park Sidings to Waterloo to carry route indicator of outward service". 03 is the code for either London Bridge to Guildford via Mitcham Jn. and Epsom, or Empty trains Waterloo to Wimbledon Park, so the meaning of 03 on an inward empty train remains a mystery for someone else to solve who was there at the time!

Page 55: this is clearly a Southern Railway era picture. (From Southern Carriage & Wagon Society lists, unit 1758 was withdrawn in February 1948.) Hence the 2/7/39 WTT is a suitable source for "L bar" as Charing Cross (or Cannon Street) to Gillingham via Parks Bridge & Loop Line.

The bar, single dot or double dot above a letter headcode were used mainly for route variations. On the Central and Eastern Sections, codes often had different meanings on outward and inward workings, e.g. only inward trains needed to distinguish between Charing Cross and Cannon Street. Empty trains fitted for letter headcodes carried a blank indicator, and a bar was used over 2-digit codes to indicate an empty train.

A curiosity in the WTT headcode listing is the occurrence of both spellings 'Park's Bridge' and 'Parks Bridge' on consecutive pages!

Page 59: No 3148 is a 4 Cor, not 4 Buf. Headcode 54 (from the 30/6/52 WTT) is Victoria and Ore (not via Eastbourne), and the train appears to be arriving off the direct spur from Stonecross Jct. This routing was used on Summer Saturdays and Bank Holidays when trains from Ore and Eastbourne were duplicated. From the few Special Notices I have, the stock is unusual, rather than a combination of 6 Pul/Pan units.

I have details of Basingstoke's N15x duties from the Sunday EWN dated 14 June 1953. All on passenger trains.

Duty 231. a.m.
9.48 Basingstoke-Eastleigh 10.31 [9. 5 from Reading] p.m.
12.48 Eastleigh-Basingstoke {1.39/1.45} -Waterloo 3. 0
5.54 Waterloo-Basingstoke 7.14
7.44 Basingstoke-Southampton T 8.45 (Shunt own train)
10.20 Southampton Ctl.-Basingstoke 11.28

3 sets of Basingstoke men on duty 8.18, 1.24, 6.54
Duty 234 am.
8.30 Basingstoke-Portsmouth & S. 9.57

Crude enlargement of the image at image at Blackheath referred to by Tony Wardle.

> **Between Sunbury and Shepperton**
>
> Berthing of empty trains on the up Line: - Between certain times on Kempton Park Race days, as advised in the Race Notice, the up line between Shepperton and Sunbury will be used for berthing empty race trains, and the down line between these stations will be worked as a single line in accordance with the Electric Train tablet Regulations.
>
> The Divisional Superintendent will provide a Relief Signalman at Sunbury to hand the tablet to the Driver of a down train and to receive the tablet from the Driver of an up train.
>
> The Inspector in charge of the race arrangements at Sunbury must send a man to Shepperton by the last down train prior to the opening of the line to ordinary traffic to see that the up line is clear of trains; and this man must take a written authority from the Inspector to the Station Master at Shepperton for the resumption of ordinary double line working, and travel with the first up train run over the proper line.

Light Engine Fratton Yard-Basingstoke or as ordered

Basingstoke men on duty 7. 0."

Continuing on with the N15X from **Jim Gosden**.

*"What a superb photograph in **ST8** of **N15X No 2329 Stephenson** in Southern Malachite Green at Waterloo. They were truly well proportioned, handsome locomotives. What a pity that they were never the equal of the King Arthurs.*

A C Perryman who as an apprentice at Brighton worked on the Baltic tanks referred to them in his book 'The Brighton Baltics' (Oakwood Press 1973) and throws some light on why this was so.

It would appear that L B Billington was impressed with the J Class 4 6 2 tank No 326 Bessborough with Walschaerts valve gear and its free running characteristics. To quote Perryman, 'So pleased was he with the success of this loco, he tried to make an even better one in the Baltics. He enlarged all his dimensions with one important omission, the diameter of the piston valves and in my opinion that was the fatal mistake."

He continues, *"The result of the small piston valves was a general reluctance to run really fast. The drivers soon found they could 'pull them up' to 35% for level or uphill running but it was necessary to advance to 40% to get any increase in speed downhill. This seems to point to the inability of the small 10 inch valves to get rid of the large volume of exhaust steam produced by the 22 inch bore cylinders which also had a longer stroke than Bessborough - further aggravating the position. This reluctance to run fast was also woefully apparent when the engines were eventually converted to tender engines."*

He goes on to say that *"The (Perryman) is building a 5-inch gauge version of a Brighton Baltic and as a result of his thoughts on the improvements that should have been done on the Baltics, he was making his piston valves 1 inch in diameter which at a scale of 1 inch to 1 foot equates to 12 inches full size. Since this is a big increase on 10 inches diameter, at the time of the rebuilding any thoughts of improving the locos may have led to the need for new cylinder castings which could explain why the rebuilds were not altered.*
With regard to the Baltics travelling to Eastleigh. Perryman confirms O J Morris's account that the cab had to be dismantled but the connecting rods were also removed and the crossheads blocked with wood in their central position. He says, '...these spare parts accompanied the engine to and fro in a 10 ton open wagon. On return to Brighton they had to revisit the shops to have these items reinstated'. So Morris may be wrong in saying that the engines were worked in steam to Eastleigh. Perryman has a picture in his book of a Baltic tank at Eastleigh in the condition he describes."

We now refer to **ST7** from **Nick Stanbury**.

"ST7 is another interesting issue. You might find the following comments of interest for 'From the Footplate':

1. A detailed article by J N Faulkner on the special traffic arrangements on Kempton Park race days, including those for the single line working, appeared in the 'Railway Magazine' in April 1955 (pages 275/276) and can be accessed at https://sremg.org.uk/RlyMag/EasterMondayAtKemptonPark.pdf

Of further interest are the relevant Southern Railway instructions from the Western Appendix of 1934: These and other Appendix instructions (amended if and when necessary) remained in force until the Southern Region issued replacement Appendices in 1960. Interestingly, there are no special Kempton Park instruction in the Western Section Appendix for 1960 or subsequently, confirming my belief that the single line working arrangements were discontinued in the late-1950s, after the Faulkner article. Perhaps someone else can add to this?"

We would agree entirely with Nick that this is certainly something unusual, reference to the 1955 article providing some further additional information.

The race specials are divided between the Kingston and Richmond routes. The early specials deposit their Waterloo passengers at Kempton Park whence they continue to Sunbury at which location they cross over the up line and return to Waterloo for more passengers.

The first race normally starts at 2.00pm therefore after about 12.30pm there is no point in sending further trains back to Waterloo and it is at this point that berthing takes place on the branch. Single line working now commences using permanently installed tablet instruments in the boxes at Sunbury and Shepperton. From now on when a race train arrives at Sunbury it crosses to the up line and runs 'wrong direction' to a point about half a mile from Shepperton where it is stabled. Each successive train follows a similar action stopping close to the preceding train except where it is necessary to maintain access for occupation crossings.

The regular half-hourly Shepperton service is maintained on the remaining line using the electric train tablet.

In the afternoon further empty trains will arrive from Strawberry Hill and Durnsford Road and these will run through to Shepperton where they berth either in the sidings or cross to the up line behind the already stabled units. In total the number of units can occupy a distance of one and half miles. Most of these trains are of eight-cars, mostly a pair of four-cars but some including 2-cars. A stand-by engine is provided at Sunbury in case of emergency.

In due course the trains move forward to collect returning race goers whilst the regular half hourly branch service (hopefully) continues to operate. Normal working' resumes around 6.00pm.

If any reader has any images of these stabled units we would be most interested. We can also ask 'were there any other Southern lines where this occurred?' - we know of course in later years new electric stock was stabled between Haywards Heath and Ardingly.

SOUTHERN TIMES

It would be interesting to know if the existing signalling was used or drivers instructed to pass some signals at danger. Might not even Pilotman working have been easier?

Finally from Nick, 2. The 'boys in boaters' watching (and perhaps participating in) the Civil Defence exercise at Kemp Town (page 43) are certainly not from Roedean – then and now an 'all-gels' school! They are almost certainly pupils at Brighton College – then all-boys but now co-ed, and sited almost next door to the erstwhile station – and doubtless drawn to the event (which was on a Sunday), whether invited or not.

Our good friend **Pawel Nowak** has added to the Kempton Park situation with his own comments. *"Upon a recent visit to Truro I was able to acquire a copy of the 1990 'Middleton Press' title on the 'Kingston and Hounslow Loops including the Shepperton Branch'. (This is not one we have in the ST library - Ed.) Lo and behold, Illustration 56 taken at Sunbury shows a 4-SUB coming off the reversible line in the Up direction, while another unit stands on the normal Up line in the distance. A few pages on, Illustration No 58 taken at Upper Halliford shows a long line of EMUs waiting to go back into service at the close of racing. This photo dated 1954."*

Next from **Ian Gordon**, and **ST8**. *"Page 62 of ST8 is Charing Cross Station and not London Bridge. This narrows down the possible service to Charing Cross to Gillingham or Maidstone West. Also Page 70 top picture, the tower indicates that the scene is on the river bridge outside Cannon Street Station, not London Bridge."*

Finally for this issue a useful note from **Michael Frackiewicz** going back to Issue 7. *"I am responding to a request for details relating to an article that appears in **Issue 7** on **page 61**. The caption describes the picture as showing the frames of a dismantled 4-4-0 with a request for more details.*

*'On closer inspection of the photo, I believe the frames belong to a 2-6-0 **No 31850** which, following a heavy shunt with another N Class, was cut up/ dismantled at the back of Redhill Shed. I have this information from Paul Abbott the Shed Master and a number of photos of the dismembering of the engine taken by one of the drivers based there at the time. I've attached some of the photos I have of the dismantling. I hope this helps.*

Michael also kindly sent two images to illustrate his comments. *Both J Jacob*

Perhaps the editor is permitted his own addition as well. Just as the recent book on the Tavern cars had gone to print (see blatent advert on the inside front cover of this issue), Andrew at the office located this view of a rebuilt restaurant car at Exeter on 13 October 1962. (Not that there was any more room in the book anyway.) The locomotive is a very grimy No 34030 *Watersmeet,* the coach and its associated kitchen car waiting to be attached to an up arrival for London. *A E Bennett / Transport Treasury*